ATM Technology
for Broadband Telecommunications Networks

ATM Technology
for Broadband Telecommunications Networks

Abhijit S. Pandya
Florida Atlantic University
Boca Raton, Florida

Ercan Sen
Siemens Telecom Networks
Boca Raton, Florida

CRC Press
Boca Raton   London   New York   Washington, D.C.

Library of Congress Cataloging-in-Publication Data

Pandya, Abhijit S.
   ATM technology for broadband telecommunication networks / by
Abhijit S. Pandya and Ercan Sen.
      p. cm.
   Includes bibliographical references and index.
   ISBN 0-8493-3139-0 (alk. paper)
   1. Asynchronous transfer mode. 2. Broadband communication
systems. I. Sen, Ercan. II. Title.
TK5105.35.P36 1998
004.6′6—dc21   98-34477
                                                                          CIP

Preface

In the last few years, the Asynchronous Transfer Mode (ATM) data transmission technique has been promoted as the foundation for the next generation of telecommunication infrastructure worldwide. Fueled by the tremendous expansion of Internet service as well as the strong support by the U.S. government for development of the Information Super Highway in the U.S., a tremendous amount of effort is being spent by the standardization organizations such as ITU-T in Europe and the ATM Forum in the U.S. as well as by the telecommunication industry and the scientific community. As with any other hot and trendy technology development, ATM technology is experiencing certain media hype and being presented as the ultimate technology for the telecommunication infrastructure of the 21st century. Our main objective in this book is to go beyond the media hype and provide the reader with a more objective and realistic view of ATM technology, its potentials and realistic deployment strategies.

Transition to ATM technology will be an evolutionary not a revolutionary process due to the tremendous amount of investment already poured into the current telecommunication infrastructure worldwide. Certainly we cannot afford to switch to a fully ATM-based telecommunication environment overnight. It is our opinion that a significant ATM-based telecommunication infrastructure will take at least a decade to develop.

As history indicates, technology alone cannot drive its success unless there is a significant demand for its use and the solution the new technology is providing is economical. Unless ATM technology satisfies these criteria, its fate may not be any different than some of the technologies developed in the last decade, such as N-ISDN.

We identify two critical features of ATM technology which will allow it to create necessary market conditions for a large-scale deployment: its ability to allow inter-working of today's several incompatible communication technologies and to provide a high-speed, high bandwidth backbone telecommunication network to satisfy the tremendous demand created by the Internet service. Eventually, we can expect that ATM technology will replace currently deployed multiple telecommunication technologies.

The original intent of this book is to provide telecommunication professionals with a compressive view of ATM technology so that they can make a more informed evaluation of the technology in developing their ATM deployment strategies. The book can also be used as a textbook for an advanced undergraduate or graduate course on ATM technology as well as an introduction to broadband telecommunication networks.

About the Authors

Dr. A. S. Pandya is an associate professor at the Computer Science Department, Florida Atlantic University. He has published over 75 papers and book chapters, and a number of books in the areas of neural networks and learning paradigms. This includes a text published by CRC Press and IEEE Press entitled, "Pattern Recognition using Neural Networks in C++." He consults for several companies including IBM, Motorola, Coulter Industries, and the U.S. Patent Office. He received his undergraduate education at I.I.T., Bombay. He also has a M.S. and a Ph.D. in Computer Science from Syracuse University, New York.

Dr. Ercan Sen is a Sr. Product Manager at Siemens Telecom Networks, Boca Raton, Florida. Previously, he has held various engineering staff and supervisory positions at Siemens Telecom Networks. He has over seventeen years of experience in the field of software design for Electronic Telephone Switching systems. He received the B.S. degree in electrical engineering from the Middle East Technical University, Ankara, Turkey, in 1981, and the M.E. and Ph.D. degrees in computer engineering from Florida Atlantic University, Boca Raton, Florida, in 1990 and 1996, respectively. His research interests include neural network applications and high-speed telecommunication networks. He is a member of IEEE since 1982.

Acknowledgments

We are grateful to Dr. Mohammad Ilyas, Chair of Computer Science and Engineering at Florida Atlantic University for his support and encouragement during the preparation of this manuscript. We also had strong encouragement from Dr. Borko Furht, Dr. Neal Coulter, Dr. Ravi Shankar, and Dr. Sam Hsu of the CSE department of FAU, which is appreciated.

We would like to thank Prof. R. Sudhakar, Prof. P. Neelakanta and Prof. J.A.S. Kelso at FAU, who have shared their knowledge and expertise over the last several years. We would like to thank our friend Bill McLean for providing his support to this project.

I (E.S) would like to thank my colleagues at Siemens Telecom Networks for their support, encouragement and insightful discussions during the writing of the manuscript.

It has been truly a pleasure to work with our editor, Jerry Papke, of CRC press in the production of the book. We are indebted to him for encouraging us to embark on to this project, his patience, understanding, and his constructive inputs on many parts of this book. We are most grateful to Mimi Williams for her meticulous effort. We are thankful to the graphics artists Dawn Boyd and Jonathan Pennell for the cover design.

I (A.P.) would like to thank my father Dr. S.P. Pandya for providing the inspiration through his own illustrious career to continue to write books. I (A.P.) would like to thank my wife Bhairavi for her understanding, patience, and help in the preparation of illustrations and the manuscript.

I (E.S.) would like to express my utmost gratitude to my wife Phoungchi and my son Justin for their sacrifice, patience and understanding which made it possible for me to allocate my time for writing the manuscript.

This book is dedicated to:

my children Rajsee and Divya.
(A.P.)

my wife Phuongchi and my son Justin.
(E.S.)

Table of Contents

Chapter 1

INTRODUCTION

At the present time, the Asynchronous Transfer Mode (ATM) data transmission technique [McDysan 1995, Stallings 1995] is being promoted as the foundation for the next generation global telecommunication infrastructure. Fueled by the tremendous expansion of the Internet service as well as the strong support by the U.S government for the development of the Information Super Highway in the U.S., a tremendous amount of effort is being spent by the standardization organizations such as ATM Forum in the U.S. and ITU-T in Europe, as well as the telecommunications industry and the scientific community.

We identify two critical features of ATM technology which will likely create necessary market conditions for a large-scale deployment: its ability to allow inter-working of various incompatible telecommunication technologies which exist today, and its ability to provide a high-speed, high-bandwidth backbone telecommunication network to satisfy the tremendous demand created by the Internet service. Eventually, we expect that the ATM technology will replace currently deployed multiple telecommunication technologies in a decade.

In an ATM network environment, the messages generated by various sources (data, voice, video, etc.) are divided into fixed-length (53-octet) packets called ATM cells and transmitted over a transmission path using statistical multiplexing technique. Due to statistical multiplexing of ATM cells, optimization of routing of these cells in an ATM switching network has been a significant challenge for the scientists.

Two main contributors to this challenge are the high-speed operation of ATM networks and the multiplexing of cells with different statistical characteristics generated by various sources. These two factors require a high-speed and highly adaptive routing controller for ATM switching networks in order to quickly adapt to fast changing cell traffic and maintain an acceptable level of Quality of Service (QOS).

I. OUTLINE OF THE BOOK

The book is organized as follows. Chapter 1 provides a market analysis for ATM technology. Chapter 2 introduces readers to the basic ATM concepts. In Chapter 3, we elaborate on the capabilities of ATM technology to

1

offer new or improved services. In Chapter 4, we discuss integration of various access technologies into ATM networks. Chapter 5 introduces readers to ATM protocols. Chapter 6 elaborates on various bandwidth allocation schemes defined for the ATM networks. In Chapter 7, we investigate switching architectures and elaborate on queuing and routing models, buffer management and broadcast and multicast requirements. Chapter 8 deals with traffic management issues. In Chapter 9, we discuss major ATM interworking standards such as LAN Emulation and IP Interworking. Chapter 10 provides a computer simulation environment that can be used for performance analysis of ATM networks. In Chapter 11, we provide an insight to ongoing ATM standardization activities currently taking place within the main standardization bodies such as ATM Forum and ITU-T.

II. CHARACTERIZATION OF BROADBAND TELECOMMUNICATIONS MARKET

The North American telecom market is the world's most competitive market. Understanding current and future market and customer trends and the drivers that cause these trends have to be the key components of a successful product design process.

The degree of competitiveness of the market is a key factor that should be considered in product planning and it is directly related to the following factors:

1) Scale of investment required

2) Scale of market size and reward

3) Regulatory conditions

4) Customer driven factors

5) Market trends

6) Influence of computing on telecommunications market

A. Scale of Investment

The smaller the scale, i.e., up to U.S. 50 million dollars, the more players, especially the small existing companies or start-up companies, will try to enter the market. These small size players are able to tap into the venture capital and stock markets to finance their product development. However, in order to attract this external capital, these companies have to come up with a very attractive product that addresses a specific need and it has to be highly market and customer oriented. Typically, these products either target a specific present need ignored by other players (a niche market) or they bring an innovative approach based on a new technology or existing technology to achieve a significant cost-benefit value for a particular market segment.

These small players operate under great pressure from the venture capitalists and the stock market that provide the necessary capital and from other players to prove themselves in the marketplace. As a result of such pressure, they tend to operate very efficiently and are able to bring their products to the market in a very short period of time. Probably the best way to describe this mode of operation would be that they operate on survival instincts. As the scale of investment requirement becomes very large, only a few large companies with enough resources are able to participate. Thus the degree of competitiveness goes down.

At this end of the spectrum, the players tend to operate in a cooperative manner forced upon them by the market and regulations. Both the customers and regulators want to make sure that there is some degree of competition so that one company does not become a dominant player and move towards a monopolistic behavior. These few players move toward a balance in terms of market share based on the market forces as well as the capacity of the players.

Due to the decreased competitiveness, the products tend not to be cost efficient or customer oriented. In other words, they do not operate in the mode of survival instinct. In a small company operating on survival instinct, there is an intense focus on a product and the resources of the company are allocated very efficiently according to the intensity of the focus. However, in a large company, the focus is not as intense as it should be and the product has to compete with many other products addressing different markets, in some cases competing for the same market segment, for company resources.

The failure of one product does not significantly impact the financial operation of a large company. Thus, the lack of focus causes an inability to adapt quickly to the changes in a highly competitive marketplace. It is very typical in large companies to see a product roadmap covering a 4 to 5 year period even for a very competitive market that is constantly changing due to intense competition. On the other hand, a small start-up company does not have 4 to 5 years to survive in the marketplace. These small companies are able to change or enhance the product quickly as the market evolves to a new state.

As an example, consider the LAN/WAN networking and related access transport market for a highly competitive market versus the public broadband backbone networking market for the other end of the scale as a less competitive market.

B. Scale of Market Size and Reward

The larger the size of the current and potential market the more players want to get in and differentiate themselves from the competitors in terms of price and performance via current technology or new technology, filling a particular gap between interrelated product segments such as the gap between the private LAN networking world and the public telecom world. Additionally, the size of the market also determines the scale of the reward. Hence, the bigger the reward, the more players are willing to compete for the reward and intrinsically increase the competitiveness of the market.

C. Regulatory Conditions

As the market becomes less regulated by the government, it tends to become more competitive. Deregulation provides new opportunities for more vendors, thus leading to a variety of products.

D. Customer Driven Factors

The customer's perception of a company and its product determines the success of a product in the marketplace. Understanding the customer's perception and the selection process used by customers in the product purchase decision is a key factor to succeed in the marketplace.

A potential customer goes through a process of product selection typically based on multi-level selection criteria. It is very important to understand this selection process and make sure that a product we are designing matches this selection criteria. The degree of match to this selection process ultimately determines the success of the product. The multi-level selection criteria can be described as:

Level-1 Selection Criteria:

Bandwidth demand: This is the main factor that causes a customer to look for new transport equipment to meet the expanding bandwidth demand. Particularly, the business customers (mid to large size corporations and universities) tend to exhaust their available bandwidth within a 2 to 3 year period due to expansion of computer networking and software applications requiring higher bandwidth per user. Rapid expansion of Internet/Intranet based communication is moving the traffic from local LAN clusters to LAN backbone, WAN and VPN networks. These customers tend to react to bandwidth exhaust rather quickly as the pressure within the organization builds up to satisfy the new bandwidth capacity demand. The bandwidth demand is probably the most critical factor for a customer to decide to upgrade its network to meet current and future bandwidth demand. Therefore, a customer in this situation will look for systems that will provide sufficient bandwidth for present and future bandwidth demand. For the future bandwidth demand, scalability, flexibility, stability of the technology and future-proof characteristics are the most important factors. Customers tend to prefer incremental upgrades to meet the future bandwidth demand instead of replacing the whole system, i.e., desire to avoid large capital investment. This will be the main selection criteria to narrow down the possible vendor/equipment list.

Bandwidth Cost: After identifying a set of equipment which fit the demand in terms of bandwidth capacity, the next step is to further narrow the list based on bandwidth cost. Investment in networking equipment directly affects the financial bottom line of a corporation, i.e., the cost of bandwidth contributes to the operating expenses of a corporation.

Level-2 Selection Criteria:

Once the bandwidth capacity and bandwidth cost factors are used to narrow the list of possible equipment, a secondary selection criteria is used to narrow the list further. The secondary selection criteria include factors such as ease of use, i.e., network management, flexible fault-tolerant features to increase reliability, quality of technical/customer support, reliability on on-time delivery and future prospects of the potential vendors.

In summary, it is important to consider both market and customer driven factors when designing a telecom product and define the features according to the relative importance as mentioned above.

E. Market Trends

The market also exhibits certain variant and non-variant trends. The variant trends usually have a limited life-cycle. This category includes technology, social behavior, regulatory rules, economic conditions, etc. Only a few non-variant trends stand the test of time and they are not influenced by the transient changes in the marketplace. For the telecom market, the two most important non-variant trends are the demand for higher bandwidth and demand for lower bandwidth cost. These two non-variant trends are intrinsically coupled to the similar trends in the computing market.

F. Influence of Computing on the Telecommunications Market

Today, computing is becoming more network oriented. As the computing power increases and the cost of computing decreases, these non-variant trends also require similar trends in the networking (communication) area. Even if we move from the PC-based distributed computing where the computer power is concentrated at the edge of the network, toward a Network-Centric (NC) computing environment where the computing power is concentrated at the core, the non-variant trends still hold.

Shifting the computing power from edge to core will achieve a greater economy of scale in terms of computing cost. However, this model will require higher bandwidth connection between the very powerful servers and low-cost NC computers to compensate for the loss of local computing power at the edge.

III. MARKET ANALYSIS FOR ATM TECHNOLOGY

A broadband communication network can be characterized by its transport and switching elements. The transport element provides the physical communication paths between the users and switching nodes and the physical paths between these switching nodes. The switching nodes allow setting up on-demand or permanent communication links between the end users of the network.

The transport element of the network is typically divided into three segments: access (local), metro, and long-haul transport segments. The access transport segment describes the edge of the network where end users are connected to the network. The access segment typically covers the first 1 to 25 miles of the network from the edge. The metro segment covers city-wide connections which extend to a 25 to 100 mile range. The long-haul segment describes the backbone of the network which covers distances over 100 miles to connect switching nodes. Figure 1-1 illustrates such network segmentation.

The key market drivers for today's telecommunication networks can be characterized as:

- Demand for higher bandwidth
- Demand for lower bandwidth cost
- Scalable network architecture to meet future bandwidth demand
- Future-proof solutions to protect investment
- Flexible bandwidth allocation for various applications
- Bridging the technology gap between private and public networks. Hence, increasing competition to lower bandwidth cost.

There are two major battlegrounds currently underway for market domination in the transport and switching segments of the broadband telecommunication market. The battleground in the transport market has to do with how to transport combined data/voice/video traffic across the broadband networks.

One of the most intense battles is underway in the access transport segment for both residential and business (enterprise/campus) segments. In the residential segment, the technology battle is among cable-TV transport, telephony transport and wireless transport technologies. The key issue is how to transport video that requires over 1-6 Mbit/s data transmission rate cost effectively without requiring massive infrastructure build-up, i.e., by using existing infrastructure.

The other key issue is how to combine (bundle) voice, data, and video (multi-media) services under one transport system to create significant economies of scale for a very cost effective solution. On the business side, i.e., private networking that includes LAN, WAN and VPN, there are multiple alternative technologies that are being promoted by different players in order to achieve market acceptance (dominance). Particularly in the LAN area, technologies such as Fast Ethernet, Gigabit Ethernet, Fiber Channel and ATM are being promoted.

On the other hand, in the core and long-haul transport segments that are considered as public segments, significant cost reduction needs to be achieved in order to support the required bandwidth demand created in the access segment. The question is how to create additional capacity in the core and long-haul segment using a fiber-based platform.

7

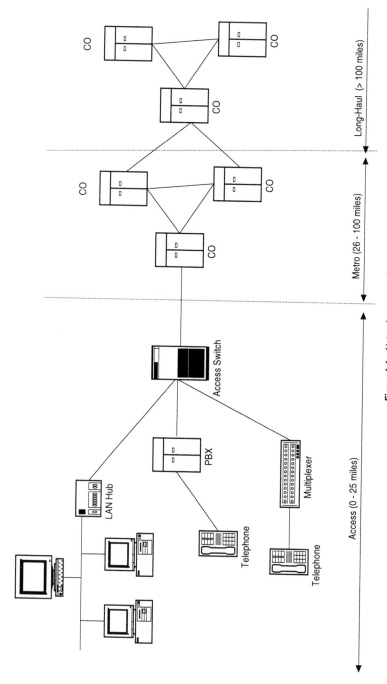

Figure 1-1 Network segmentation.

8

There are two possible alternatives: add additional fiber to the network or increase the capacity of the existing fiber network via higher speed Time Division Multiplexing (TDM)-based Synchronous Optical NETwork (SONET) technology or Dense Wavelength Division Multiplexing (DWDM) technology. We will explain these technologies in more detail later.

In the area of switching, two concepts are the prime candidates for market domination: cell-based ATM switching and frame-based IP switching. The major question is which one is more efficient than the other. The answer depends on the particular environment or the traffic type.

If the originating traffic is IP based, i.e., most of the LAN and Internet traffic, and the traffic can tolerate variable delay, then the answer is IP switching is more efficient. On the other hand, if the traffic is highly delay sensitive such as voice and we have to deal with the existing PCM-based telephony infrastructure, then the cell-based ATM is the answer. Although, it is possible to carry IP traffic over ATM network (IP over ATM), due to the additional layer of segmentation and reassembly of IP frames, as well as the additional overhead of ATM cell headers, it is less efficient than transporting IP frames as a whole.

There other aspects which separate these two switching technologies: for example, ATM offers better QOS guarantees since it is a connection oriented switching technology. There are attempts to offset this aspect of ATM by introducing reservation schemes such as RSVP for the IP technology. Since the current Internet and LAN transport is frame based, the IP switching provides compatibility with these major traffic sources. On both fronts, however, there are attempts to emulate the features of the other side to offset the advantages. All these major competitive forces have been the result of the 1996 Telecommunication Act approved by the U.S. Congress. In the following sections, we analyze each market segment in more detail.

A. Residential Market

The key market dominance battle in this segment is between the cable operators and the telephone operators. Each camp has a similar strategy: trying to protect its own market while attempting to enter another's market. For example, the cable operators are trying to offer telephony and Internet service besides their original TV service using their cable infrastructure. On the other hand, the telephone operators are trying to offer video service using their telephone infrastructure. The two competitive factors involved here are the sense of survival instinct to protect one's own market and the lure of additional profit to be made by entering a new market segment.

At the present time, there is no evidence of any major gain by either side. The telephone companies are currently carrying the majority of the Internet traffic as well as the traditional telephone traffic. The second analog line or N-ISDN line demand for Internet access has been a major growth and revenue source for the telephone companies. However, the increased Internet traffic is putting pressure on the current telephone network. The telephone companies are forced to add additional capacity without the possibility of recovering the cost of additional equipment due to regulatory restriction on basic residential telephone service. Until the telephone companies find a way to impose higher access fees for Internet access, their dilemma will continue.

Some major telephone companies such as Bell Atlantic and U.S. West acquired cable companies to get into the cable-TV business. However, the telephone companies are prohibited from offering cable-TV service in their own telephone market. Therefore, they are not able to achieve economies of scale by combining the telephone and TV broadcast service on the same infrastructure.

Following the passage of the 1996 Telecom Act, there was a flurry of merger and acquisition activity among the cable-TV operators and the telephone operators. However, recently, the cable-TV industry has been under significant pressure from the Direct Satellite Broadcast (DBS) industry. Hence, the cable-TV operators have not been able to spend large amounts of capital due to price pressure. As a result, some merger initiatives fell through and some acquisitions by the telephone operators are faltering. For example, U.S. West recently announced separating its cable operation from its telephone operation.

Looking at the near-term prospects, the cable-TV operators are poised to gain some customers from the telephone operators for the Internet service by offering high-bandwidth Internet access at a 10 to 30 Mbit/s rate at a affordable price around the U.S.$40 range. However, due to the common shared medium nature of the cable, as more users access the system, the performance of the system will degrade and the promised bandwidth will not be available to the subscribers. On the other hand, the telephone operators are counting on the new technologies such as xDSL (ADSL, HDSL, VDSL) to increase the bandwidth to subscriber loop, thus able to provide both on demand video and high-speed Internet access. The ADSL technology is likely to be deployed in 1998 by major telephone operators.

B. Business Market

There is an intense battle currently underway in the business market segment for market domination. The business market segment is the most revenue rich segment that is willing to adopt new solutions to gain a competitive edge through an advance communication platform. The telecommunication equipment manufacturers are promoting various communication technologies to gain market domination. The two key technologies are Ethernet and ATM technologies. Until the introduction of ATM technology, the business market had been mainly dominated by the frame based Ethernet technology.

The major players like CISCO and 3COM have been consistently improving their Ethernet-based LAN technology in order to maintain their dominant market position. They have been very successful in maintaining market share by bringing switched Ethernet, Fast Ethernet and most recently Gigabit Ethernet enhancements to the Ethernet technology. With these improvements in the Ethernet technology they were able to meet the ever increasing bandwidth demand and at the same time maintain lower bandwidth cost compared to other competing technologies such as ATM.

Most recently, there has been a major shift in the market towards a more unified networking to handle data, voice and video traffic within the same network. For example, businesses would like to consolidate their voice, data and video on the single networking platform to achieve significant cost reductions and ease of network manageability. The Ethernet works very well for pure data traffic. However, it is not well suited to carry delay sensitive voice and video traffic. The Ethernet technology does not provide bandwidth allocation (management) and guarantee of stringent QOS requirements for specific traffic types.

There are some attempts to bring these key features to Ethernet through reservation schemes such as RSVP. However, these are not native to the original concept of Ethernet technology which was intended for a shared and statistically very bursty pure data traffic environment. Currently, the only key advantage the Ethernet has over ATM is cost of bandwidth. It is possible to offer Ethernet connection under $100 per user port today.

The ATM technology has been designed from the beginning to allow dynamic and flexible bandwidth allocation and provides required QOS for different traffic types for each user. The penetration of ATM technology in the business segment has been slow due to the cost factor. However, the cost of ATM per user port is coming down rapidly. For example, the cost of 25 Mbit/s connection has come down to just under $1000. The major factor for the reduction in cost is the siliconization (putting functions in the Integrated Circuits) of most of the ATM layer functions with standard interfaces.

C. Bandwidth Cost Trend

The bandwidth cost is expected to continually decline as the bandwidth capacity increases as a result of using technologies such as Dense Wavelength Division Multiplexing (DWDM) and the expansion of the fiber network using high quality Polarization Mode Dispersion-(PMD) free fiber cable. Figure 1-2 provides a projection of such a trend relative to current bandwidth cost for access (local) transport segment. A similar trend is also valid for metro and long-haul segments as well. As seen in Figure 1-2, significant bandwidth cost reduction can be achieved as a function of capacity increase. However, the important factor is the relative capacity increase with respect to the associated cost involved. The rate of capacity increase has to be greater than the rate of cost increase.

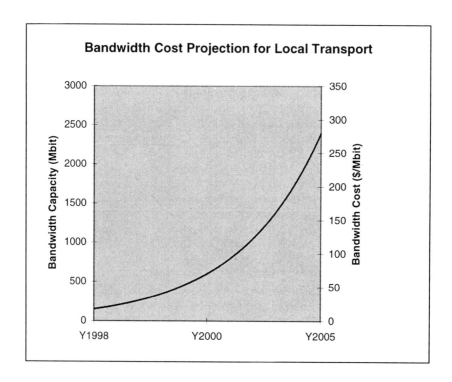

Figure 1-2 Bandwidth cost projection for local transport.

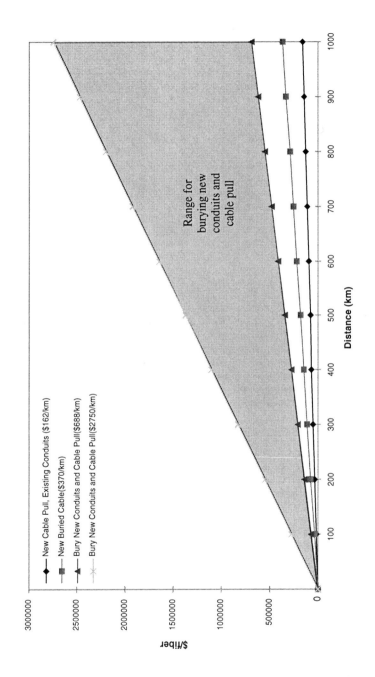

When we consider DWDM and installation of new fiber for capacity increase as two possible alternatives, the DWDM is much more favorable to installing new fiber cable. Particularly in distances less than 400 miles where no DWDM regenerators are required, the DWDM provides much higher cost reduction than the new fiber installation. This is mainly because the rate of capacity increase is much higher relative to rate of DWDM equipment cost.

In the case of fiber installation, it is very difficult to achieve cost reduction due to the very labor intensive nature of the process. As it can be seen in Figure 1-3, the cost of fiber installation linearly increases with distance. Hence, it is obvious to assume that significant bandwidth cost reduction will mainly come from the utilization of DWDM technology.

Figure 1-4 illustrates implementation of the DWDM-based transport system for 500 km and 1000 km distances. Based on the current technology, it is necessary to regenerate each wavelength at every 500 km distance due to the accumulative noise effect from the optical line amplifiers. It is expected that further advances in optical technology will push the maximum distance without signal regeneration to longer distances, hence reducing the cost of DWDM further for the long-haul transport segment.

D. Market Opportunities

The two emerging technologies DWDM- and ATM-based VP (Virtual Path) present very significant opportunities in the transport market today. These two technologies are capable of collapsing the present SONET multiplexing hierarchy to achieve a more efficient and cost effective transport medium.

For example, it is possible to use both the DWDM and ATM technologies to collapse the current SONET hierarchy into a leaner, more efficient and cost effective transport system. In the long-term, advances in optical transmission and optical switching will play a pivotal role in terms of increasing the wavelength density and making it possible to implement a fully optical communication network.

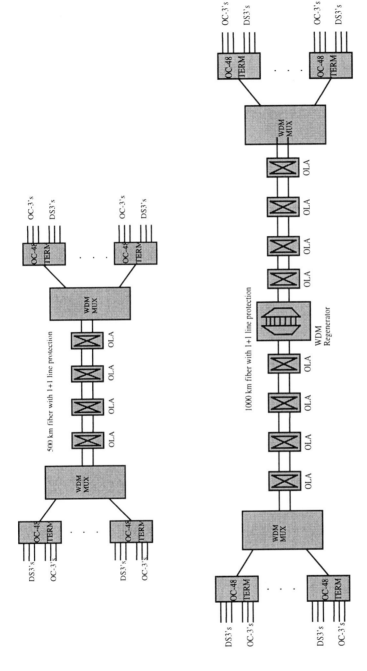

Figure 1-4 DWDM implementation (a) for 500 km distance, (b) for 1000 km distance.

15

Chapter 2

BASIC ATM CONCEPTS

Asynchronous transfer mode (ATM) technology promises the integration of the diverse services of voice, video, image and data. ATM is a technology that allows transmission and switching of fixed-length packets known as cells through an ATM switching network asynchronously. The fixed-length ATM cell structure defines the foundation of the ATM technology. This chapter provides a high-level overview of ATM concepts.

I. ATM OVERVIEW

There has been a growing interest in the development of integrated multi-service enterprise and service provider networks, which consolidate voice, video, imaging, and computer data traffic into a single network. The driving forces for this interest are the rapid rate of growth in non-voice traffic, the emergence of new strategic multimedia applications, and the significant cost savings which can be realized from network consolidation. A common direction for both enterprises and service providers is the evolution towards ATM networking.

High-speed networks refer to networks with a link capacity of 100 Mbits/s and above. They are also high capacity networks. If a large number of users are sharing the capacity, their behavior is not fundamentally different from lower speed networks. However, in the case of a small number of high-bandwidth users the latency may start to dominate performance. Thus it is critical to reduce or hide end-to-end latency. Table 2-1 lists various alternatives for high speed networking.

A principal attribute of ATM is that it is equally suitable for departmental and campus local area networks (LAN), metropolitan area networks (MAN) and wide area networks (WAN). In order to attain the end-to-end integration of broadband services, interoperability of LANs, MANs and WANs is a key requirement. For the first time this one technology is positioned to provide an end-to-end solution for the user.

Table 2-1

Telecommunications (WAN or Wide Area Networks)	LAN/MAN (Local/Metropolitan Area Networks)	I/O Architectures
B-ISDN (Broadband ISDN)	FDDI (Fiber Distributed Data Interface)	ANSI Fiber Channel
SMDS (Switched Multi-Megabit Services)	IEEE 802.6 DQDB (Distributed-Queue Dual-Bus)	HIPPI (HIgh Performance Parallel Interface)
SONET (Synchronous Optical NETwork)	—	—

Each application has different requirements, for example, some are more sensitive than others. Data related applications can tolerate delay, but not loss. On the other hand, audio and video can tolerate loss, but not delay. Table 2-2 provides information on various applications, their traffic characteristics and service requirements.

One of the most significant promises of asynchronous transfer mode (ATM) technology is the integration of diversified services such as voice, video, image, and data. Some of the salient features of ATM can be listed as follows:

1) It integrates Voice, Video and Data

2) ATM uses short, fixed length packets called cells

3) It is a best effort delivery system

4) It provides bandwidth on demand

5) It is a connection oriented technology, i.e., every cell with the same source and destination travels over the same route

6) It has the potential to remove performance bottlenecks in today's LANs and WANs

7) It knits local and wide-area networks and services into a seamless whole

8) It is possible to bill the customer on a per-cell basis

9) ATM is scalable i.e., it works at different speeds and on a variety of media

10) ATM architecture has an open-ended growth path, since it is not locked to any physical medium or speed

There are certain disadvantages of ATM:

1) Overhead of cell header (5 bytes per cell)

2) Complex mechanisms for achieving Quality of Service

3) Congestion may cause cell losses

Application	Traffic Characteristics	Service Requirements
Electronic mail	Individual messages size: < 1Kbyte to several Mbytes Typically short connection times	Maximum delay: several minutes or more low bandwidth no end-to-end data loss allowed
Voice (PCM coded)	Constant bit rate size: 64 Kbps long connection times uniform traffic pattern	Maximum delay < 200ms guaranteed/fixed bandwidth low cell loss
Browsing databases (including graphics)	Size per picture: 100 Kbytes to 10 Mbytes bursty traffic	Maximum delay: typically 10 to 100 ms
Real-time video (MPEG)	Variable bit rate size: 128 Kbps 1.5 Mbps bursty traffic pattern	Low average delay, low delay variation, low cell loss ($< 10^{-10}$) required
Traffic between workstations/PCs and centralized data bases	Peak rates of 10 Mbps or more short connection times, typically bursty traffic	Maximum delay: 1 to 100 ms depends on the application

ATM technology can also be viewed as the next stage in the evolution of packet switching and circuit-switching technologies. Packet switching as the name suggests is a method in which long messages are subdivided into short packets and then transmitted through a communications network. In case of circuit switching a dedicated communications path is established through one or more intermediate switching nodes. Digital data are sent as a continuous stream of bits and the data rate is guaranteed. Delay in this case is limited to propagation time.

ATM combines the two switching techniques and takes advantage of the desirable characteristics of both techniques. The statistical multiplexing feature from packet switching allows efficient use of the available bandwidth while the connection oriented feature from circuit-switching provides predictable cell transmission delay. Another strong reason for selection of fixed-length for the ATM cell and the connection-oriented path selection is to be able to implement the cell-switching at the hardware level in order to achieve the high-speed data transmission requirement.

ATM allows multiple logical connections to be multiplexed over a single physical interface. The information flow on each logical connection is organized into cells. Many data (as opposed to voice or video) applications are bursty in nature and can more efficiently be handled with some sort of packet-switching approach. It is important to note that ATM does not provide for error-control or flow-control mechanisms.

Asynchronous, in our context, refers to the ability of the ATM network to send only the data associated with a connection when there are actual live data that need to be sent. In contrast, *Synchronous* Transfer Mode (STM) networks consisting of channels require a continuous stream of data and when the channel is idle a special bit pattern, called *idle cell,* must be sent in every time slot representing the channel.

There are several disadvantages of the synchronous approach. First, it does not provide a flexible interface for meeting a variety of needs. At the high data rates offered by ATM, there could be a wide variety of applications supported by ATM involving many different data rates. One or two fixed-rate channel types in this case would not provide a structure that can easily accommodate this requirement. Another aspect of the inflexibility of the synchronous approach is that it does not lend itself to rate adaptation which is essential for channels containing hundreds of megabits per second.

Figure 2-1 shows the overall hierarchy of function in an ATM-based network. This hierarchy is seen from the point of view of the internal network functions needed to support ATM as well as the user-network functions.

Figure 2-1 illustrates the architecture of the ATM model. The ATM layer consists of virtual channel and virtual path levels as illustrated in Figure 2-2. These two components are discussed in detail in the next subsection. Below that is the physical layer as shown in Figure 2-1.

Based on functional differences the physical layer can be divided into three levels (see Figure 2-2):

Transmission path level: This level consists of network elements that assemble and disassemble the payload of a transmission system. The payload in the case of end-to-end communication is the end-user information. The payload for user-to-network communication may be signaling information. Cell delineation and header error-control functions are required at the endpoints of each transmission path.

Digital section: This section of the physical layer consists of network elements that assemble and disassemble a continuous bit or byte stream. This refers to the exchanges or signal transfer points in a network that are involved in switching data streams.

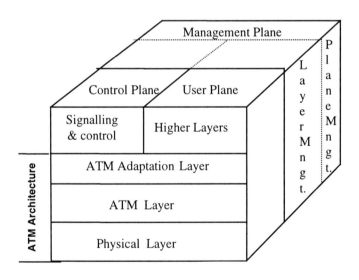

Figure 2-1 B-ISDN Protocol Reference Model.

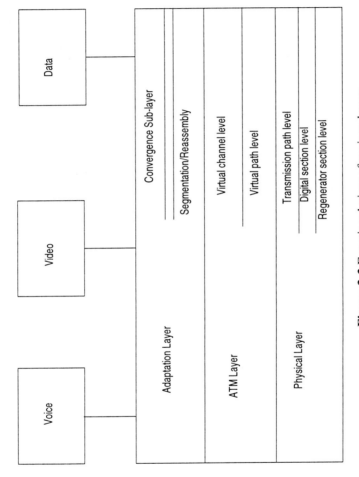

Figure 2-2 Functional view of various layers.

23

Regenerator section level: We can regard this level as a portion of the digital section. An example of this level is a repeater that is used to simply regenerate the digital signal along a transmission path that is too long to be used without such regeneration. In this case there is no switching involved.

II. BASIC DEFINITIONS FOR ATM CONCEPTS

A.   Virtual Path and Virtual Circuit Concepts

Logical connections in ATM are referred to as **virtual channels** and they are the basic units of switching. A virtual channel is set up between two end users through the network, and a variable-rate, full duplex flow of fixed-size cells is exchanged over the connection. Virtual channels are also used for user-network exchange, i.e., control signaling and   network-network exchange, i.e., network management and routing.

A bundle of virtual channels that have the same endpoints is referred to as a **virtual path**. As a result, all the cells flowing over all of the virtual channels in a single virtual path are switched together.

The Virtual Path (VP) and Virtual Circuit (VC) concepts are the basic building blocks for identifying a physical transmission path to carry ATM cells associated with a particular call. The relationship between VP, VC and the transmission path is illustrated in Figure 2-3.

As mentioned earlier, ATM is based on connection-oriented switching. That is, the transmission path is determined and the necessary routing relationship at the intermediate ATM switching nodes is established at call setup time.

In the ATM network environment, routing is localized to adjacent nodes along the transmission path. Each node only has the knowledge of the next node along the path and this partial routing information is contained in the VP and VC fields of the cell header. These two fields are updated at each node as the cell travels along its transmission path.

The VP identifies a particular physical transmission link between two adjacent nodes while the VC identifies a   logical sub-bandwidth in that particular transmission link. The available physical bandwidth on a particular VP is dynamically partitioned among several VCs. A VP can be dynamically partitioned into 65,535 VCs.

Today there is a trend in high speed networking in which the cost related to the control of the network is an increasingly higher proportion of the overall network cost [Burg 1991]. The virtual path technique was developed

24

to contain the control cost by grouping connections sharing common paths through the network into a single unit. As a result the network management actions have to be applied to a small number of groups of connections instead of a large number of individual connections.

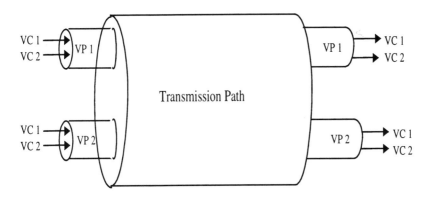

Figure 2-3 ATM Virtual Path and Virtual Channel Concept.

Several advantages can be listed for the use of virtual paths:

Simplified network architecture: Network transport functions can be separated into those related to an individual logical connection (virtual channel) and those related to a group of logical connections (virtual path).

Increased network performance and reliability: The network deals with fewer, aggregated entities.

Reduced processing and short connection set up time: Much of the work is done when the virtual path is set up. By reserving capacity on a virtual path connection in anticipation of later call arrivals, new virtual channel connections can be established by executing simple control functions at the endpoints of the virtual path connection; no call processing is required at transit nodes. Thus, the addition of new virtual channels to an existing virtual path involves minimal processing.

Enhanced network services: The virtual path is used internal to the network but is also visible to the end user. Thus, the user may define closed user groups or closed networks of virtual-channel bundles.

A virtual channel connection (VCC) provides end-to-end transfer of ATM cells between ATM users (usually the ATM adaptation layer). The endpoints may be end users, network entities, or an end user and a network entity. Each endpoint associates a unique virtual channel identifier (VCI) with each VCC. The two endpoints may employ different VCIs for the same VCC.

In addition, within the network, there may be a number of points at which virtual channels are switched, and at those points the VCI may be changed. Thus, a VCC consists of a concatenation of one or more virtual channel links, with the VCI remaining constant for the extent of the VC link and changing at the VC switch points.

Between an endpoint and a VC switch point, or between two VC switch points, a virtual path connection (VPC) provides a route for all VC links that share the two VPC endpoints.

B. Interfaces

ATM networking technology defines a number of standardized interfaces which will allow the interconnection of switches and ATM-connected routers and workstations from multiple vendors. User-to-Network Interfaces (UNIs) are responsible for providing interconnections from end-systems to an ATM switch. Network-Node Interface (NNI) connects ATM switches. Figure 2-4 shows a network with various interfaces.

ATM can be used both for public networks as well as private networks. Thus there are three distinct forms of UNIs defined primarily based on physical reach:

Public UNI: This will typically be used to interconnect an ATM user with an ATM switch deployed in a public service provider network. These UNIs must be capable of spanning long distances since they must connect users to switches in public central offices.

Private UNI: This will typically be used to interconnect an ATM user with an ATM switch that is managed as part of the same corporate network. For example users connecting to a private ATM in a corporation or university will use the Private UNI. They can use limited distance technologies since the private switching equipment can often be located in the same room as the user device.

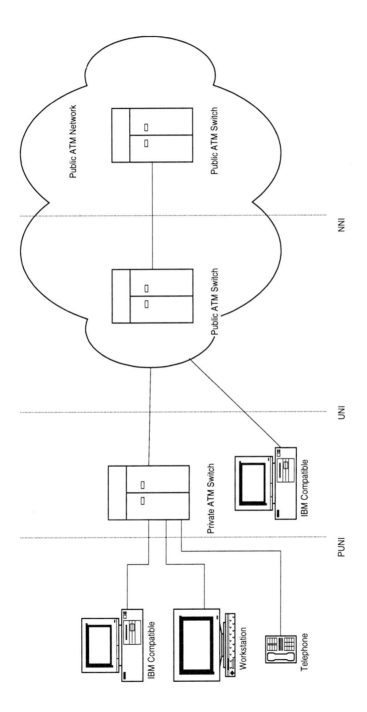

Figure 2-4 An ATM network with various interfaces.

Residential UNI: Typically used for interconnection of residential end-systems such as set-top units and the ATM service. This may be considered a varient of the Public UNI since it is deployed in a public setting. However the signaling, transmission speed, and the physical media requirements may be quite different in this case.

Similarly, there are three types of NNIs for connecting ATM switches.

Private NNI (PNNI): Typically deployed as a part of a private ATM.

Public NNI: Typically used for providing the interconnection of ATM switches within a single service provider.

Broadband Inter-carrier Interface (B-ICI): Typically used to trunk multiple service providers.

III.   ATM CELL STRUCTURE

The basic ATM cell structure is composed of a 5-octet header and 48-octet payload (user data) as shown in Figure 2-5. The incoming user data bit stream at the ATM User-Network Interface (UNI) is divided into 48-octet cells with a 5-octet header added to each cell before the cells are transmitted into the ATM network.

The header contains the necessary information to allow the ATM network to deliver a cell to its destination. The Generic Flow Control (GCF) field is used for bandwidth policing to make sure that users obey the predetermined bandwidth allocation negotiated at the time of call set up. The GCF field is only present at the UNI but not at the Network-Node Interface (NNI) since ATM relies on end-to-end flow control at the user level rather at the network level.

The Virtual Path Identifier (VPI) and Virtual Channel Identifier (VCI) fields hold the address information for an ATM switching node to determine the route for the cell. Once the route is determined, the ATM node updates these fields for the next ATM node in the path. The Payload Type (PT) field identifies the cell as containing user data, signaling data or maintenance information.

The Cell Loss Priority (CLP) field indicates whether the cell is a high-priority or a low-priority cell. This field is used for congestion control purposes and allows the congestion control mechanism to identify low-priority cells to be discarded during congestion periods.

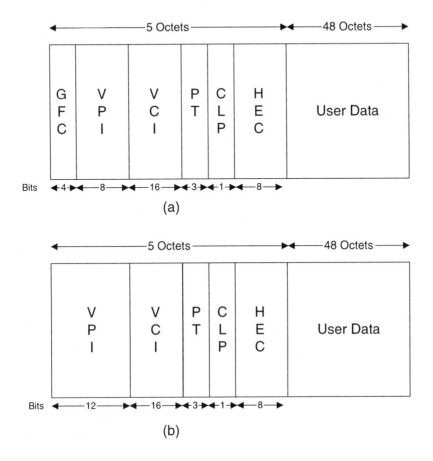

Figure 2-5 ATM cell format at (a) UNI and (b) NNI.

The last header field is the Header Error Check (HEC) field which allows error detection and error correction in the header. No error checking or correction applies to payload data as it passes through the network. They are left to higher level protocols at the end nodes since the ATM switch has no knowledge of the user data content.

IV. SIGNALING IN ATM NETWORKS

A. Introduction

There are two important developments that will shape telecommunications networks in the near future. One is a significant increase in the demand for bandwidth, resulting especially from the explosive growth in volumes of data. The other is the convergence of overlay data and switched voice networks to form a broadband ISDN network.

There are a variety of technologies available to adapt to these changes; however, ATM is the most promising. Designed for a wide range of traffic types, including voice, video and data, ATM is best qualified for broadband ISDN, and it can scale rapidly to meet the growing demand for bandwidth.

Until recently, ATM networks have been run using permanent virtual circuits (PVCs). End customers of any future broadband ISDN network, however, will want bandwidth on demand. This only becomes possible with the introduction of switched virtual circuits (SVCs).

Permanent virtual connections: For permanent virtual connections (virtual path or virtual channel), the service contract is defined at the user end-points and along the route that has been selected for the connection using the provisioning capabilities of the network vendor.

Switched virtual connections: Switched virtual connections enable the network to allocate bandwidth on demand. End users make their service request to the network using functions provided by the end system for accessing the ATM user-to-network interface (UNI).

ATM networks with SVC capabilities have advantages over today's PVC-only networks. An end customer with SVCs only pays for the connection time and the bandwidth that is used. While there will always be customers who require PVC services, SVC services will offer economies in many cases (depending on the length of time of the connection and tariff rates).

Consequently, the customer base is expected to grow rapidly. Carriers will be able to be more flexible in the way they operate the networks, and their costs will be lower since less administrative intervention will be necessary. Driven by the obvious market and administrative advantages, carriers are demanding SVC-enabled ATM networks.

30

Despite these obvious business drivers there are a number of significant technical challenges posed by SVCs, especially in the area of signaling and routing. Information for connection control, traffic management and carrier-specific service features must be transferred between exchanges using signaling. The successful integration of high speed ATM networks with the various legacy data and telephone networks will place high demands on signaling throughout the ATM network. Routing is also critical for traffic control and will play an important role in protecting against traffic overload in SVC-enabled ATM networks.

In summary, if carriers are to meet the demand for switched bandwidth services, their core and network access ATM equipment will need to be able to handle all widely used signaling interfaces while accommodating interworking between the individual variants.

This section presents the signaling systems that are currently in use. It also provides a glimpse into the complexity of future networks and the demands that this will place on signaling in general.

B. Signaling Mechanisms

We shall now discuss the signaling response for defining the procedures to be used when establishing connections over both, the interface between the end user and the ATM switch. In ATM switch networks, the user provides the information on the virtual channels to the ATM nodes.

Any connection on the ATM network is referred to as a *call* and the process where the user informs the ATM network what connections, i.e., VPCs or VCCs, need to be set up is referred to *as call control*.

ATM signaling must be capable of doing a number of things, since ATM networks have a wide range of capabilities. For example, multi-connection calls can be set up between users and services like transferring voice; video and data simultaneously may be required. During such a call connection it is important that these connections can be set up "on the fly", when needed.

For video services, one cell sent into the network should be delivered to a number of end points. This would require a point to multi point type of connection. For more complex video and voice conferencing arrangements, a multi-point to multi-point type of connection is required. Figure 2-6 shows the signaling sequence for a point-to-point connection. Figure 2-7 shows the multi-point signaling sequence.

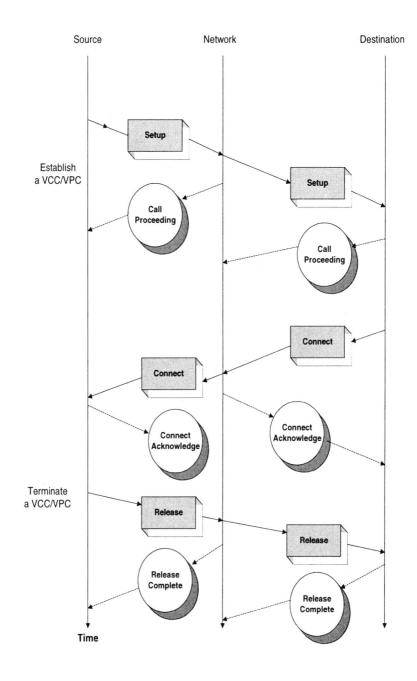

Figure 2-6 Signaling sequence for point-to-point connection.

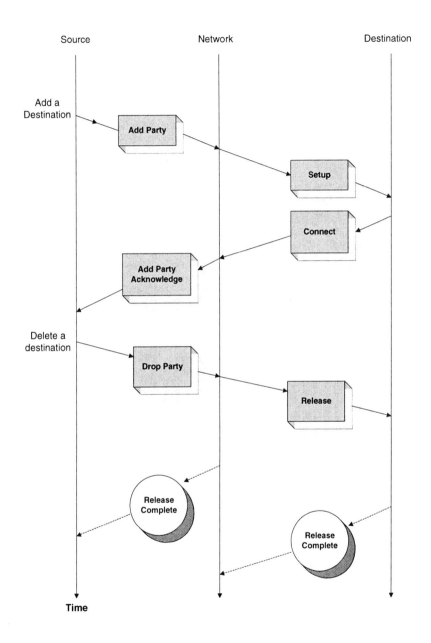

Figure 2-7 Signaling sequence for multi-point connections.

ATM signaling must be able to establish, maintain and release both the actual user virtual channel connections (VCCs) and the static virtual path connections (VPCs) in order to allow information flow across the network. It should be able to negotiate the appropriate traffic and service characteristics of a connection.

In ATM, signaling is done by cells sent from the user to the network and back again using a *logical* signaling channel. Signaling is not performed over a channel used for traffic, so it is called out-of-band signaling. Current ITU standard ATM implements call control as follows.

VPIs are assigned at the *service provision time* on a per site basis, resulting in a "permanent" way for site connectivity. Such a VC establishment mechanism is called a Permanent Virtual Circuit (PVC). However, the VCIs are set up through "dynamic" connections between users at these sites by means of a signaling protocol. This kind of real-time signaling mechanism is called a Switched Virtual Circuit (SVC).

These protocols are used by devices to set up, maintain, change and terminate connections across the ATM network. For example, the customer premise equipment (CPE) sends a message to the local network node requesting a connection to Miami. If the destination (Miami) accepts the connection set up by the ATM network, it will send a message to the originator, such as, "connection OK, use VCI = 155." VPI is predefined in this case as mentioned earlier.

The Permanent Virtual Connections (PVCs) are set up using an ATM Layer Management function on a node by node basis. In other words, segments of a PVC are established between each ATM node involved manually using the layer management function. This type of connection requires coordination between each node and manual verification of the complete path (end-to-end). However, due to the permanent nature of the connection, the manual set up overhead is acceptable.

On the other hand, the Switched Virtual Connections (SVCs) require a robust signaling protocol for call set up and release due to dynamic and on demand nature of the connections. The current ATM standard for ATM signaling is ITU-T Q2931 which is adapted by both ITU-T and the ATM Forum. The Q2931 is part of the UNI 3.1 specification. The Q.2931is an adaptation of ITU-T Q931 ISDN Narrowband Signaling protocol with enhancements to support point-to-multi-point connections. The latest version of the UNI specification (UNI 4.0) adds additional capabilities such as negotiation of QoS parameters and VPI/VCIs at the UNI interface.

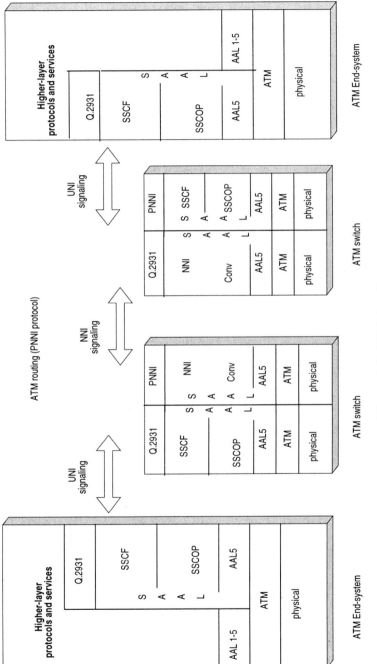

Figure 2-8 ATM signaling.

35

Figure 2-8 illustrates the signaling mechanism across an ATM network.

C. ATM Addressing

Due to the fact that different addressing schemes are used today for private and public telecommunication domains, the ATM Forum recommendation on ATM addressing involves two sets of addressing schemes, one for private and one for public ATM networks. An important element of this addressing concept is to maintain interoperability with existing networks which is one of the key objectives of ATM technology.

The ATM Forum recommends use of OSI's Network Service Access Point (NSAP) address format for identification of end-users in private ATM networks and E.164 format for the public UNI interfaces in the public domain. The structure of the E.164 ATM Addressing as recommended by the ATM Forum is shown in Figure 2-9. The upper 13 octets (AFI, E.164,HO-DSP) constitute the public domain part (network prefix).

The network prefix part contains information about the granting authority and routing hierarchy. The eight-octet E.164 part contains the addressing information currently used in the worldwide telephone network. The lower seven-octet End-System Identifier (ESI) field describes the identity of an end-user in the private network environment. For example, the ESI may represent the MAC address of a LAN workstation.

The Interim Local Management Interface (ILMI) bridges the public and private domains. When an end-station in a private LAN network wants to communicate to another end-user over the public ATM network, it uses the ILMI to register itself to the ATM switch which serves as the public UNI interface gateway for the private network. The end-station sends its MAC address as ESI to the ATM switch and receives back a fully qualified E.164 ATM address from the ATM switch. The address registration through ILMI is illustrated in Figure 2-10.

V. ATM SWITCHING CONCEPTS

One of the key concepts in ATM technology is that is a switched based technology. By providing connectivity through a switch (instead of a shared bus) several benefits are provided. We can list the benefits as follows:

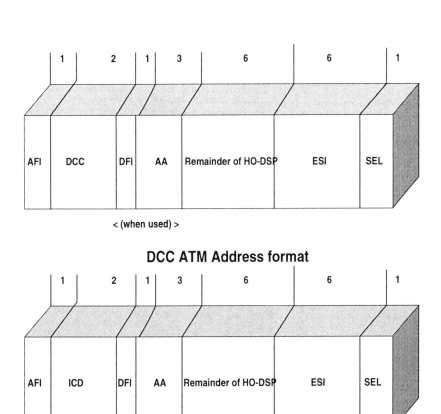

DCC ATM Address format

ICD ATM Address format

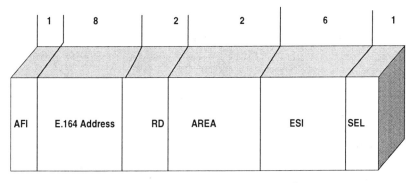

NSAP encapsulated E.164 Address format

Figure 2-9 ATM address formats.

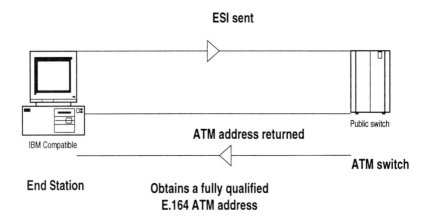

Figure 2-10 Addressing sequence.

1) Dedicated bandwidth per connection
2) Higher aggregate bandwidth
3) Well-defined connection procedures
4) Flexible access speeds

A detailed discussion on how broadband signaling fits into the architecture of an ATM network can be found in Black (1997). Black introduces the fundamentals of broadband signaling, and explains its role in ATM networks. He also introduces ISDN, B-ISDN and ATM architectures, and presents the components of ATM signaling in detail. Readers can find a good in-depth coverage of the Signaling ATM Adaptation Layer (SAAL), which is responsible for correctly transferring data on a broadband signaling link.

Figure 2-11 shows the ATM network nodes acting as virtual path switches. In this example, it can be seen that the VPIs change on a node-by-node basis; however, the VCIs do not change. The VPI translation in such cases is performed by a simple lookup table at each ATM network node. Such VP switches terminate the previously defined VP links and therefore must translate the incoming VPIs to the outgoing VPIs. On the other hand, the assignment and use of virtual channels are up to the end users, and the ATM network provides connectivity between site locations. Thus it is sometimes referred to as a virtual network. Such an arrangement is often useful for LAN-to-LAN or client-server, router-based networked applications.

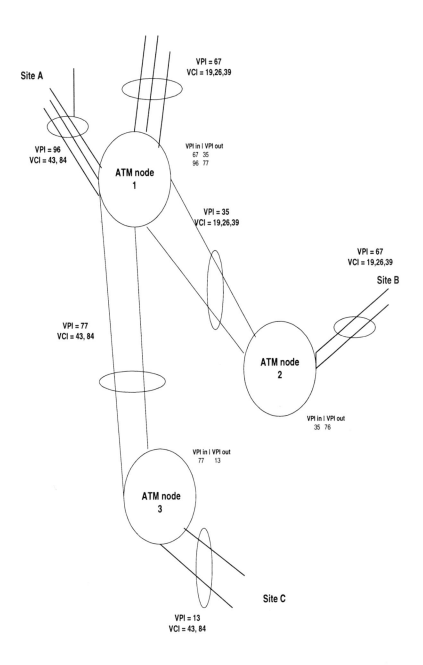

Site A

VPI = 67
VCI = 19,26,39

VPI = 96
VCI = 43, 84

VPI in | VPI out
67   35
96   77

**ATM node
1**

VPI = 35
VCI = 19,26,39

VPI = 67
VCI = 19,26,39

Site B

VPI = 77
VCI = 43, 84

**ATM node
2**

VPI in | VPI out
35   76

VPI in | VPI out
77   13

**ATM node
3**

Site C

VPI = 13
VCI = 43, 84

Figure 2-11   Pure VP switch and no VCI switch.

Although Figure 2-11 shows only a VP switch, VP switching is not the only possibility for an ATM network node. These network nodes may also switch virtual channels. However, this function must be built on top of a virtual path switch function. These simultaneous VC and VP switches terminate both VC links and VP links. VCI translation in this case is possible, again based on a table lookup, usually in a completely separate table.

It is important to note that the cell sequence -first sent, first received- is preserved during transportation. This is achieved by transporting all cells associated with a particular VPI/VCI value in a cell header along the same path. There is no dynamic routing on a cell-by-cell basis, since it would cause havoc among the cells set aside for voice/video, where elaborate bit sequencing mechanisms do not exist in end-user equipment.

A full VP and VCI switch is shown in Figure 2-12. Although it is not expected to be common because its application is not obvious, it is nevertheless allowed to have a pure VC switch in an ATM network. That is, the VP switch function exists but is just a "pass through" while the VCIs are reassigned. This device is shown in Figure 2.13. The network will have more to say about VCIs and VPIs when considering the ATM layer in more detail later.

When discussing ATM switches, it is always a good idea to keep in mind that these devices are switches, not routers. Recall that the difference in routing is that all hop-by-hop output link decisions are made based on information included in the packet header, and each packet is sent repeatedly through the routing process at each node. In switching, all hop-by-hop link decisions are made ahead of time, and resources are allocated at this point. Thereafter, no other decisions are necessary other than a quick table lookup. Header information may be kept to a minimum.

The other advantage that switch-based networks have is with the use of sequential delivery. If delivery of data is not guaranteed to be sequential, as with connectionless, router-based networks; if a destination receives packet 1 and 3 but not packet 2; the destination must wait to see if packet 2 shows up. This means that an additional delay must be built into the destination to allow for this. On the other hand, packet 2 may be "lost" in the network (a link or router fails, etc.) and never arrives. Thus the "timeout" interval must be balanced between the need to wait for an out-of-sequence packet and the need to request the sender to re-send the missing packet.

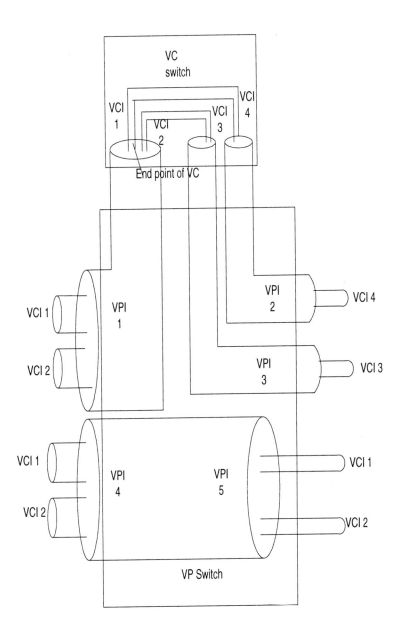

Figure 2-12 VP and VCI switch.

41

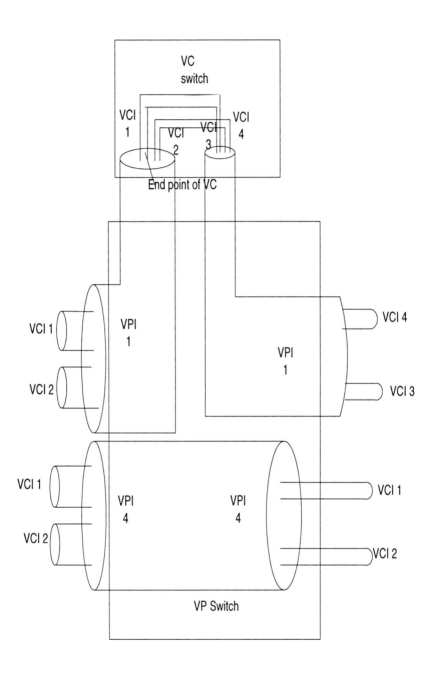

Figure 2-13 Pure VCI switch and no VP switch.

Connection-oriented, switch-based networks have no limitation. With sequential delivery guaranteed, if the destination receives packet 1 and then packet 3, packet 2 must be missing in action in the network itself. Immediate steps may be taken by the destination to correct this error condition, which may involve notifying the sender that a retransmission is needed, but not necessarily.

Thus ATM switches make decisions based on the information in a VPI/VCI table. Information in the ATM switch's VPI/VCI table is obtained in several ways, which includes operations performed for call set up.

VI.   ROUTING CELLS IN ATM NETWORKS

Routing of cells in an ATM switching network poses a challenging performance problem for the designers of these high-speed ATM switching networks. We briefly survey the research results currently available in the area of routing control in the literature.

The previous work done in this field can be classified into two categories: discrete routing control [Aras 1994, Chang 1986, Katevenis 1987, Lang 1990, Muralidar 1987] and neural network based routing control [Rauch, 1988, Brown 1989, Brown 1990, Hiramatsu 1990, Lee 1993, Troudet 1991, Sen 1995]. The discrete routing control has been around since the early 70s with the wide use of digital computers and the desire to connect these computers together in a network.

The neural network-based [Pandya 1995] routing control has been studied since the late 80s as an alternative to the discrete routing control as the knowledge on neural networks grew significantly after Hopfield [Hopfield 1985] was able to demonstrate the potential computing power of the recurrent neural networks in 1982.

In the 70s the discrete routing control was able to satisfy the need for routing in the computer communication networks which can be characterized as small, low-speed, and isolated. However, in the late 80s and early 90s, thanks to the major advances in fiber-optics transmission and integrated circuit (IC) technology, a new trend emerged to connect very high-speed computing unit through a global high-speed communication network.

This trend led naturally to the envisioning of the Information Super Highway based on the ATM technology. Such a large high-speed communication network requires a very flexible and highly adaptive routing control mechanism in order to accommodate very diverse and highly dynamic traffic patterns. Additionally, a routing controller has to be very fast in order to manage routing processes effectively in this high-speed network environment.

One can easily see that there are two major requirements for the routing controller of the emerging ATM based high-speed telecommunication network:

- high adaptability
- extremely high speed routing operation

Routing optimization can be formulated as a maximum matching problem of choosing maximum number of conflict-free cells from a connection request matrix in each time slot which is in the order of a few hundred nanoseconds in these high-speed communication networks.

There are efficient algorithms to solve such combinatorial optimization problems using conventional computing architectures. However, the cost of achieving such operations in a very short period of time using conventional discrete control based architectures is very expensive if not impossible.

Additionally, these conventional computing architectures are not very flexible to adapt smoothly to bursty incoming traffic patterns, which may adversely effect the performance of these architectures. With their dynamic and adoptive control characteristics, neural networks have the potential to satisfy the routing performance requirements of ATM switching networks.

VII. CALL SETUP

Figure 2-14 suggests in a general way the call establishment process using virtual channels and virtual paths. The process of setting up a virtual path connection is decoupled from the process of setting up an individual virtual channel connection.

The virtual path control mechanisms typically include calculating routes, allocating capacity, and storing connection state information.

For an individual virtual channel set up, control involves checking that there is a virtual path connection to the required destination node with sufficient available capacity to support the virtual channel, with the appropriate quality of service, and then storing the required state information (virtual channel/virtual path mapping).

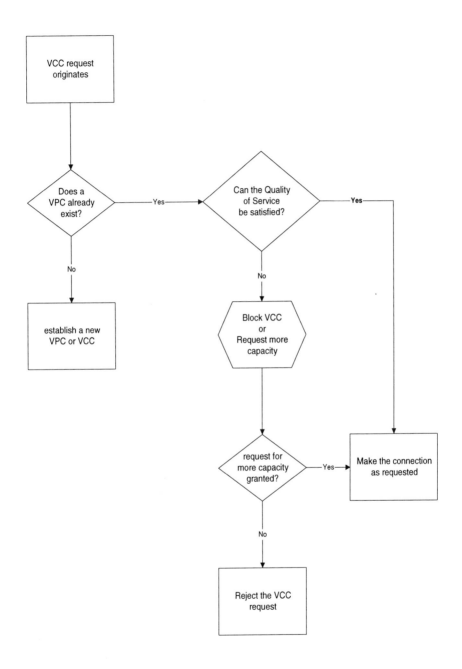

Figure 2-14 Algorithm for call setup request.

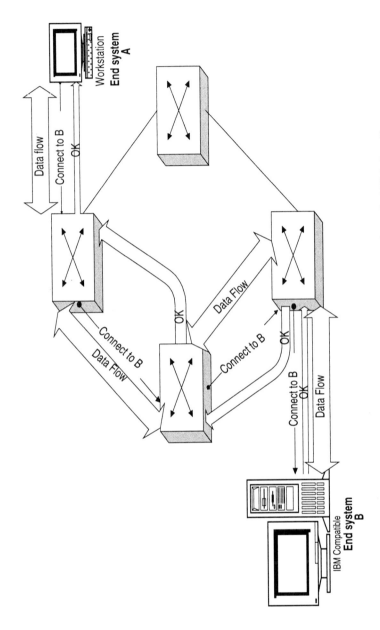

Figure 2-15 Connection set up between two end systems through ATM signaling.

Figure 2-15 illustrates how a call setup through the ATM works. First a request is signaled by end-user A to connect to end-user B. This request is routed through intermediate connections establishing a set up path. At that point the call request is accepted or denied by end-user B. If it is accepted, then the data flows along the same path. Finally the connection is torn down.

Chapter 3

SERVICES OFFERED THROUGH ATM NETWORKS

One of the key promises of ATM technology besides consolidation of current telecommunication technologies is to provide services, which are not easy or economical to provide with existing technologies. Some of the key characteristics of these services are: high bandwidth demand, long service duration, bursty traffic patterns, varying quality of service requirements, and asymmetrical bandwidth usage. In this chapter, we will investigate some of these services such as Internet, Video on Demand, Video Telephony, Distant Learning/Medicine and Telecommuting, and explain how ATM can provide these services economically.

Although some of these services we just mentioned are currently being offered through the existing public telephone network, they are too expensive to be afforded by the general public or there is not enough bandwidth available to fully utilize these services. Additionally, these fairly new services put a significant burden on the current telephony network that was originally designed for carrying traditional voice traffic.

The voice traffic can be characterized as low bandwidth (64 Kbit/s), symmetrical, delay sensitive and short service duration, i.e., average 3-minute call duration. However, the new services require bandwidth in the order of Mbit/s and the typical service time can be considered in hours. Due to the capacity limitation of the present telephony network, these services are offered either at a premium price or at a lower bandwidth with substantially lower quality.

I. INTERNET SERVICE

At the present time, Internet service is expanding rapidly. With the current growth rate it can be assumed that it will surpass the traditional voice service in the near future particularly in the United States. Especially, with the possibility of offering very cheap voice over Internet service, the Internet poses a significant challenge for telephone network operators.

49

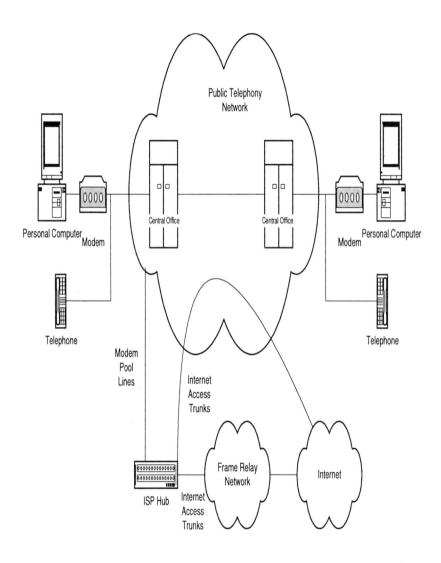

Figure 3-1 The traditional Internet Service Offering.

Telephone network operators are at a crossroad in terms of how to deal with the explosive growth of the Internet service. On the one hand, they have to invest in the infrastructure to deal with the bandwidth demand created by Internet service while figuring out a way to deal with the competitive pressure the Internet is bearing on them for the voice service. The current Internet service offering via the traditional telephone network is shown in Figure 3-1.

There are currently a few technologies being considered to deal with the Internet phenomena. The first one, which is promoted by the Cable-TV operators, is the cable modem technology. The cable modem offers bandwidth in Mbit/s range through cable-TV network. The current price of Internet service through cable modem is in the range of U.S. $40 to 50. Given the current status of Internet service via the traditional telephone network, i.e., U.S. $10 to 20 range using analog modem at rates up to 56 Kbit/sec or U.S. $40 to 100 range using the ISDN at rates 64 to 128Kbit/s, the cable modem offers significant value.

However, the cable-TV network is a shared medium. Hence, as the number of cable modem users increases, it will not be possible to sustain the same bandwidth allocation per user or the cable operators have to limit the number of users to sustain the same bandwidth level per user. Possible Internet service offerings via cable modem are shown in Figure 3-2 and Figure 3-3. In Figure 3-2, the cable company provides both the access and ISP service as a package (bundled). In Figure 3-3, the access and ISP service are separated (unbundled). The cable company charges an access fee while the ISP provider charges for the content.

The second technology solution is the Asymmetrical Digital Subscriber (ADSL) technology to increase the bandwidth to telephone subscribers up to 6 Mbit/sec range. The ADSL technology allows separation of high bandwidth Internet traffic from the low bandwidth voice traffic at the central office interface. Hence, a telephone network operator can divert the bursty Internet traffic to another network such as Frame Relay or ATM network. With this approach the telephone network operator can protect the quality of service for the traditional voice traffic.

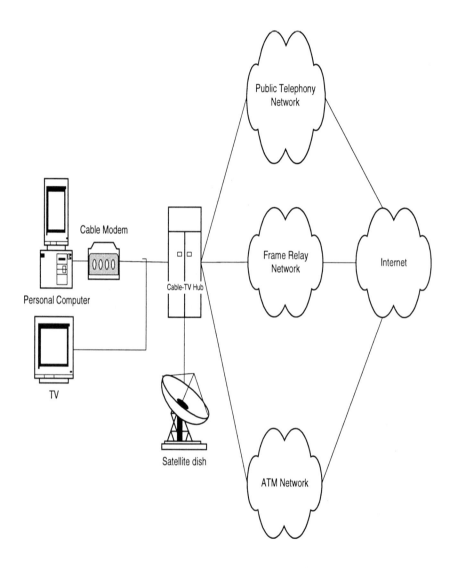

Figure 3-2 Internet service offering via cable modem in which both the access and ISP service are provided by the cable company.

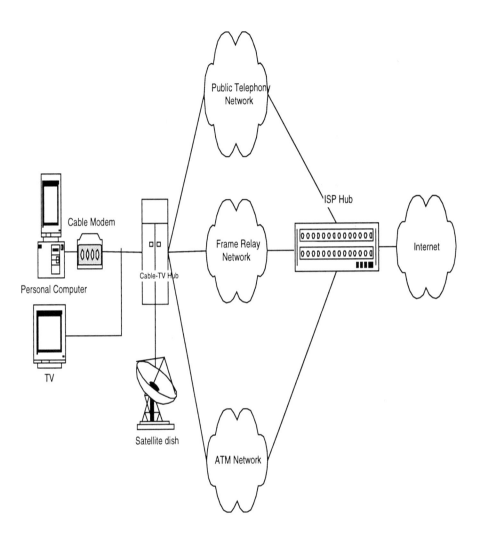

Figure 3-3 Internet service offering via cable modem in which access and ISP service are separated.

However, notice that the Frame Relay or ATM networks in which the Internet traffic are being routed may not necessarily belong to the same operator. In this case, the telephone network operators will likely charge a nominal access fee. This is really an intermediate solution, because it requires multiple overlay networks. There are two possible ADSL deployment scenarios as depicted in Figure 3-4 and Figure 3-5. Figure 3-4 shows a case where the ADSL technology is integrated into the central office. Figure 3-5, on the other hand, shows a case where an external Digital Subscriber Line Multiplexer (DSLAM) is used to separate the Internet traffic from the traditional voice traffic. In this case, no upgrade is required for the existing central office.

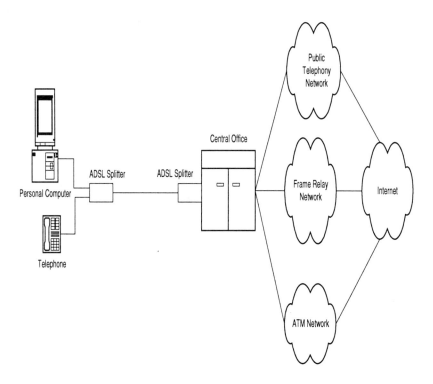

Figure 3-4 Integrated ADSL solution.

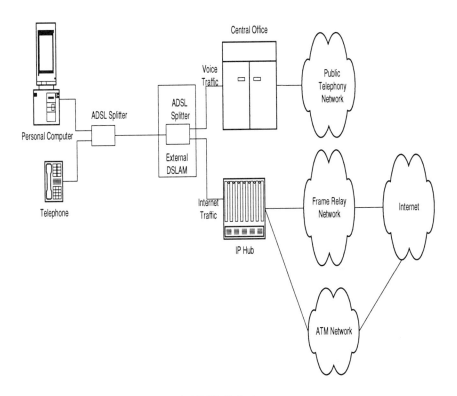

Figure 3-5 External ADSL Solution.

The third technology solution is the deployment of ATM technology all across to provide both voice and Internet service via the ATM network. The only reason for the intermediate solution using the overlay Frame Relay or ATM network is the high cost associated with the ATM technology. The transition to a final ATM-based network solution will ultimately be determined by the rate of cost reduction in ATM technology. We can envision a pure ATM-based telecommunication network across the United States when per user port cost goes down to below the U.S. $100 range. Figure 3-6 illustrates such a pure ATM-based Internet service offering.

II. VIDEO ON DEMAND

Video on Demand (VOD) will be one of the most popular and powerful service offerings through ATM technology. VOD will cause a paradigm shift in terms of how entertainment and education are provided to the public. Through VOD it will be possible to view an entertainment/new video piece whenever one wishes. In the education arena, people will be able to access educational videos stored in digital libraries. Information and entertainment retrieval will be personalized. People will no longer have to adjust their schedule according to a broadcast schedule if they wish to view a particular entertainment/news /educational program.

A successful VOD operation requires two basic infrastructure elements: high bandwidth for transmission of video content to each destination and a very efficient video server to serve many clients simultaneously with different timing.

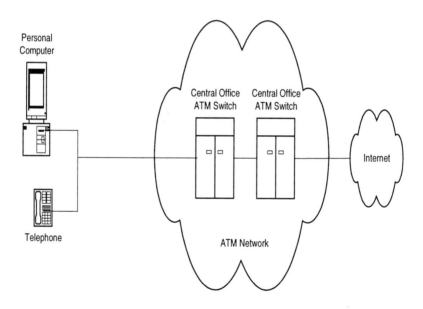

Figure 3-6 Pure ATM-based Internet service offering.

A typical video transmission carries a large amount of information and it is very bursty in nature. ATM technology is well suited for carrying video traffic very efficiently. A high quality video transmission such as a video in High Definition Television (HDTV) format requires in the order of 40 to 100 Mbit/sec bandwidth capacity without any compression.

At the present time, transmission of high quality TV broadcast through telephony network requires setting up a communication path between the source and destination at DS3 transmission rate (45 Mbit/s) using Synchronous Transfer Mode. Since the bandwidth is reserved at the peak transfer rate, it wastes bandwidth and it is very expensive. Using the ATM technology it is possible to reserve bandwidth less than the peak rate. Hence, ATM is better suited for carrying video traffic.

VOD traffic has very asymmetrical characteristics in terms of bandwidth use. Typically, the downstream traffic from the server to clients requires large bandwidth while the upstream traffic from the clients to server requires very small bandwidth for sending control information such as stop, start, pause, rewind, forward, etc. The STM technology requires equal bandwidth allocation in both directions whereas ATM allows different bandwidth allocation in both directions. Hence, from accommodating asymmetrical bandwidth for VOD service point of view, ATM is superior to STM technology.

Significant bandwidth reduction for video transmission can be achieved using compression techniques to compress the video information. The most promising video compression technique today is MPEG-2. MPEG-2 offers compression ratios over 60 times. Efforts are underway to achieve higher compression ratios and as a result, the next generation of MPEG (MPEG-3 and MPEG-4) compression techniques is currently being defined.

In order to provide a successful VOD service, the response time for serving the request has to be real-time or near real-time. Hence, the video server is a crucial piece for implementing a successful VOD service. A successful video server technology has to be capable of serving many clients viewing different segments of a video program in real-time or near real-time, i.e., the video server has to be operating in the multi-treading mode.

There are several technological and network architecture problems related to the video server that need to be solved in order to meet the challenge for a successful VOD service offering. Although ATM can provide a transport environment for point-to-point video transmission, an efficient VOD operation requires a very efficient point-to-multipoint transport environment. Another approach would be to use a cache system to maintain multiple copies of a video program in different locations closer to the clients.

III. VIDEO TELEPHONY

Video Telephony (VT) service offering will be among the favorite services provided by the ATM networks both for the general public and business community. It will probably replace the traditional voice only telecommunication form. Through VT, people will be able to communicate visually. VT will make it possible to communicate person-to-person in private or allow setting up video conferencing.

VT shares some common characteristics with VOD and it also differs from VOD in some other aspects. We will elaborate on these common and divergent characteristics in this section.

VT also requires high bandwidth to carry video signals between the communicating parties. However, the demand for high bandwidth is not as stringent as VOD. During a typical person-to-person VT session as shown in Figure 3-7, the images change relatively slowly compared to an action scene in a video movie. Also, VT does not suffer from the complexity and difficulty involved in VOD in regards to the video server.

The traffic between a person-to-person VT session is limited to two nodes and the traffic is symmetrical. Even in the case of video conferencing, as shown in Figure 3-8, in which more than two persons are involved, the number of clients is fairly small compared to the VOD. In addition, the video transmission is simultaneous, i.e., only one common image is broadcast to all conference participants. Hence, it requires a relatively simpler conference server. ATM also offers very a cost-effective solution for VT service due to the bursty nature of video traffic. It can be expected that VT service will be provided sooner than VOD due to its relative simplicity in many aspects as explained above.

IV. DISTANT LEARNING/MEDICINE

Distant Learning and Medicine service provided through ATM technology will revolutionize the way education and medicine are practiced. Geographical distance will no longer be a problem to access the best teachers and medical professionals by the public. Expert knowledge in education and medicine will be made available to anyone.

The key requirements for distant learning/medicine are high bandwidth and low bandwidth cost. ATM technology will be able to meet these requirements through efficient use of the bandwidth resources of the public networks.

High bandwidth is required because both distant learning and medicine rely on high quality video images to be carried over the network to different geographical locations. Distant medicine requires carrying high definition medical images and medical data across the network. On the other hand, distant learning requires high quality video conferencing capability involving perhaps hundred of students. Interactivity in real-time will be an important requirement for both distant learning and medicine.

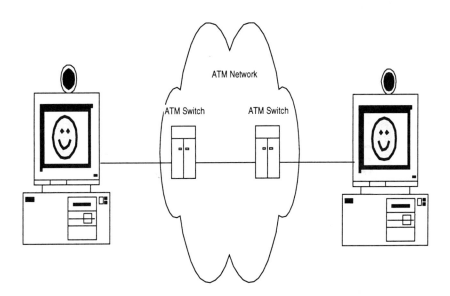

Figure 3-7 Video telephony (person-to-person).

V. TELECOMMUTING

Crowded and inadequate urban infrastructures such as roads are making working in downtown business districts almost impossible. Due to congested roads during busy hours, commuting between work and home is becoming a nightmare. Workers have to spend a considerable amount of their time (i.e., 1 to 3 hours) each day for commuting.

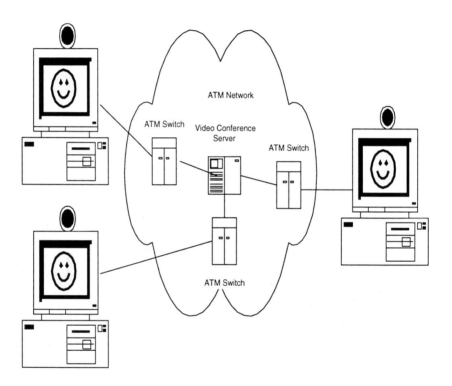

Figure 3-8 Video conferencing.

Most recently, due to advances in telecommunication, telecommuting has become an alternative to traveling each day between work and home through congested roads. Both government agencies (federal, state and local) and businesses are promoting telecommuting as a solution to reduce commuting to crowded urban areas. Recent experimentation with telecommuting is indicating significant improvements in workers' productivity and morale.

As with the other telecommunication services we described in this chapter previously, telecommuting also depends on high bandwidth and low bandwidth cost. Telecommuting will require combinations of video telephony and video conferencing, and ubiquitous access to computing resources of a company. In other words, terms of access to company resources and interaction with clients or co-workers have to be made transparent between physically being at a company's premise or telecommuting from home or a satellite location. ATM technology will certainly make telecommuting affordable to many organizations and improve the traffic conditions of the urban areas.

Chapter 4

INTEGRATION OF VARIOUS ACCESS NODE TECHNOLOGIES INTO ATM NETWORKS

Another key promise of ATM technology is to provide a platform to integrate today's various access node technologies such as wireless, Plain Old Telephony (POT), Data, Cable-TV. Hence it offers an affordable and easy access among the networks based on these diverse access technologies. Similarly, ATM is also capable of providing access among different transport technologies such as SONET/SDH and Frame Relay. Each of these access and transport technologies we have just mentioned were developed to address particular needs and were performed efficiently in their targeted communication domain. However, due to globalization of communication and the need for offering combined services such as voice, video and data, the networks built on these access and transport technologies had to be interconnected. ATM technology is well suited for this enormous interconnectivity task. In this chapter, we will elaborate on how ATM technology can be applied in each of these areas to provide interconnectivity.

I. WIRELESS NETWORKS

In recent years, wireless technology provided one of the fast growth areas for the communication industry worldwide. The key drivers for this tremendous growth in the wireless segment were:

- The promise of providing communication capability anywhere, any time, even while on the move.

- Rapid infrastructure build up at an affordable cost.

While tremendous growth in the U.S. can be mainly attributed to the first driver, in other industrialized countries and developing countries the second driver has been more dominant than the first one. For example, in most of these industrialized countries, the government has generally owned the traditional PSTN networks. Due to this monopolistic structure, the telephone service rates have been rather high compared to the U.S. However, due to the second driver, in some of these countries where the competition is allowed, private companies were able to build wireless networks to compete with the government owned traditional telephone service. In fact, Internet access in

63

these countries is still not affordable by the majority of the population. This is mainly due to dependence of Internet access on the traditional telephone access via modem connection. On the other hand, in the developing countries such as China where the problem is the lack of PSTN infrastructure, the wireless technology offers an attractive solution for rapid communication infrastructure build up.

Although the current wireless technology offers mobility and rapid and affordable communication infrastructure build up, it lacks the necessary bandwidth to be used in broadband applications such as data and video transport to user. However, there are new developments to expand use of wireless technology into broadband applications such as wireless LAN based on ATM protocol.

One of the most recent developments in the communication area is the packetized voice transport over data networks and the Internet using protocols such as voice over IP (VOI) or voice telephony over ATM (VTOA). The packetized voice allows suppression of silence periods of natural speech, hence reducing the bandwidth requirement as well as allowing sharing of the same physical channel by multiple voice connections, i.e., statistical multiplexing.

In traditional wireless networks, digitized voice samples in Pulse Code Modulation (PCM) format are carried over 64 Kbps DS0 channels. However, by use of ATM technology in wireless networks as shown in Figure 4-1, the packetized voice can be transported more efficiently from base stations to a central office location. At the central office, the voice can be converted to DS0 format to be carried over traditional PSTN networks or the packetized voice can be carried over Public Data Networks (PDN) such as Frame Relay, ATM and the Internet networks. Carrying voice over PDNs offers a more cost-effective alternative. Additionally, more voice calls can be carried out on the same physical link between base stations and central office. Advantages of using ATM between base stations and central office is two-fold:

- Higher concentration on the physical links between base stations and central office,

- Interoperability with PDNs to allow cost-effective voice transport alternatives.

The interoperability feature of ATM technology with PSTN and other data networks will be further discussed in Section 4.2 and Section 4.3, respectively.

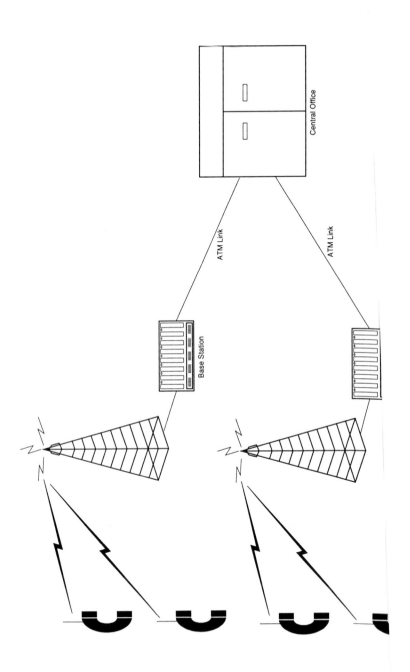

Central Office

ATM Link

ATM Link

Base Station

II. PUBLIC TELEPHONE SERVICE NETWORKS

The traditional PSTNs served well in the past to carry voice-only traffic. However, as the traffic pattern shifts from voice to data due to explosive growth experienced in Internet access, the typical telephone access is no longer a viable solution.

Although it make sense from the technology perspective to switch to a packetized transport mechanism for voice, data and video over a single universal network such as ATM, it is not economically possible to abandon PSTN networks due to the large installed base. The migration from PSTN to ATM networks will take an evolutionary path and will occur over time. This has been a general trend in the telecommunication field from the beginning. Migration from an existing technology to a new one always took place gradually.

The most probable scenario for the transition from PSTN to ATM (or any other packet transport technology such as IP) is to add the capability to separate high speed and bursty data or video traffic from the PSTN networks at the user interface using access technologies such as ADSL as shown in Figure 4-2. Once the data or video traffic is separated it can be routed to an ATM network. In this case, while the PSTN network handles the traditional voice traffic, the ATM network handles the packetized traffic (data and video).

The key factors for continuing to use the PSTN network for the traditional voice traffic instead of immediately moving to ATM network are:

- Cost
- Availability

The cost factor is the most obvious reason to maintain the PSTN network for sometime in the future. In most cases, the operators of the PSTN network have already recovered their investments on PSTN equipment. Hence, the only cost associated with the traditional voice service over the PSTN network is maintenance and administration related expenses. On the other hand, to provide the same level of service and geographical coverage by implementing an ATM network would require a very large initial investment.

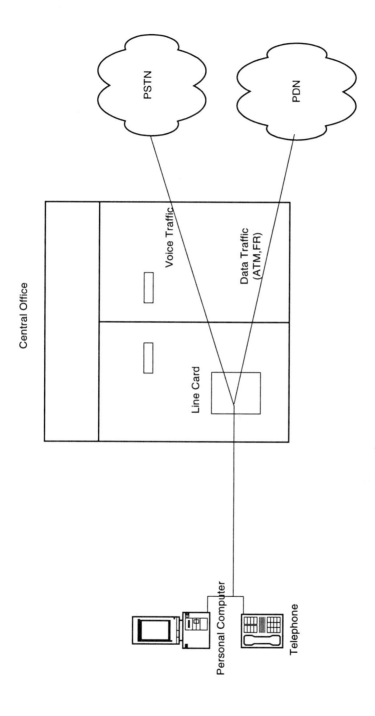

Figure 4-2 Separation of high-speed video/data traffic from PSTN network.

Particularly in the U.S., operators of the PSTN network offer a wide range of value-added voice services such as custom calling features (call waiting, call forwarding, etc.), Centrex and EKTS features for business customers. Implementation of such features for ATM networks will take enormous amount of time and effort. Hence the availability of such calling features for voice service is an important factor in maintaining the current PSTN network for sometime into the future. However, there is no doubt that we will reach a crossover point when ATM networks can provide these sophisticated voice services cost effectively.

The manufacturers of the traditional PSTN equipment already are considering various approaches to prolong the life cycle of their equipment. These approaches involve integration of ATM capability into their equipment or providing adjunct ATM capability for handling high bandwidth data and video traffic.

III. DATA NETWORKS

At present, the data networks (public or private) are segmented into various sub-networks such as Frame-Relay, SMDS, SNA, X.25, etc. This segmentation can be attributed to the development of several data transport technologies to address very specific needs. In most cases, these sub-networks are not compatible. Hence, internetworking between these networks is not possible or very costly.

Due to globalization of business and increasing dependence on information and computing, a need to interconnect these sub-networks has arisen recently. ATM technology has been developed to particularly address this internetworking problem. In other words, the enabling of internetworking of existing data networks is one of the key promises of ATM technology. Today, ATM technology is already being used successfully to interconnect various LAN technologies over Wide Area Network (WAN) applications.

Initially, LAN technologies such as Token-Ring and various Ethernet-based technologies were designed to operate in isolation in a particular business and university campus environment. In other words, a LAN would be dedicated to a particular workgroup, which tightly interact within the workgroup with little or no interaction with other workgroups. For example, a workgroup may represent a single department in a corporation or university campus. A LAN would then provide a medium to share information among the members of the workgroup.

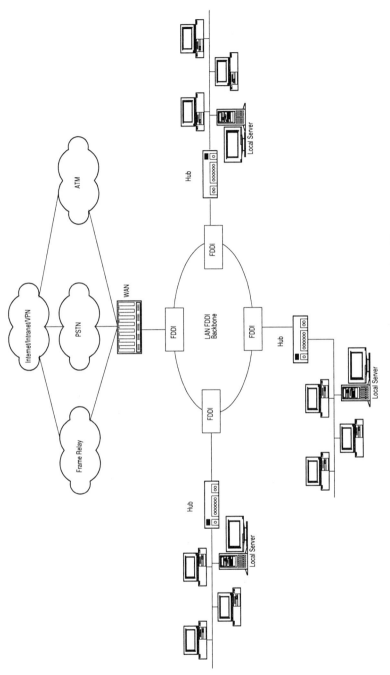

Figure 4-3 A Typical LAN FDDI Backbone.

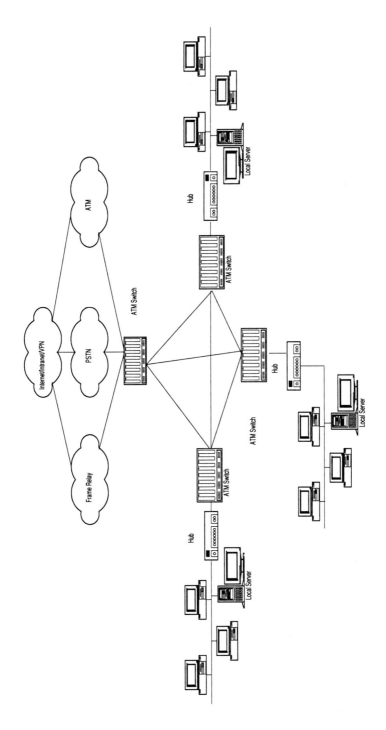

Figure 4-4 Typical LAN ATM Backbone.

The need to access global information resources and to share that information on a global basis are the cornerstones of the information age. As a result, it became necessary to interconnect these isolated LANs. As an internetworking technology, ATM will play a critical role in interconnecting these LAN clusters as part of a campus, MAN (Metropolitan Area Network), WAN, Internet/Intranet, PN/VPN (Private Network/Virtual Private Network) application.

We next look at the internetworking application scenarios for the ATM technology in the context of LAN-based networks. We first look at the business/campus setting. In this case, various LAN clusters based on the same or different LAN technologies are interconnected, typically via an FDDI (Fiber Distributed Data Interface)-based LAN backbone ring as shown in Figure 4-3. As the bandwidth demand increases beyond 100 Mbps for the LAN backbone network, ATM is expected to replace the currently used FDDI technology as illustrated in Figure 4-4. As the ATM workgroup switches become more affordable it is expected that ATM will also be used as a LAN application to connect LAN users.

IV. CABLE-TV NETWORKS

Prior to the passage of the Telecom Act'96 in February 1996 by the U.S. Congress, cable-TV networks were limited to providing TV service to the general public. One of the key objectives of the Telecom Act'96 was to break the monopoly in local telephone service provided by the RBOCs (Regional Bell Operating Companies) and other telephone operators in a particular telephone service area. There are currently three access solutions available to provide competitive telephone service. These are wireless network, cable-TV network, and unbundling of PSTN access through telephone wire connection.

Although the cable-TV network operators were very enthusiastic about getting into telephone service as a new business opportunity following the passage of Telecom Act'96, their initial enthusiasm has been dampened by the realization that it will require a significant amount of investment to offer traditional telephony service over a cable-TV network and be very difficult to offer cheaper telephone service than the PSTN operators, particularly in the U.S.

However, these cable-TV operators soon realized that they could offer high speed Internet access, instead of telephony service, to cable-TV subscribers using the large amount of bandwidth available in their cable-TV networks. At the present time, cable-TV network operators are already offering fast Internet access while telephone operators are still struggling with the type of technology to use (such as ADSL) to provide fast Internet access to their subscribers. Eventually, as VOI and VTOA become mature technologies and widely used, then it will be possible for cable-TV operators to provide very cost-effective telephony service to their customers.

We next look at how ATM technology can be used in cable-TV networks to provide fast Internet access. The cable-TV networks are currently capable of providing 10 to 30Mbps down-link capacity to the cable user. However, the up-link from cable user to cable ISP is still problematic. In most cases, the up-link requires a modem connection over PSTN to cable ISP with low bandwidth capacity. This situation still maintains the monopoly of PSTN operators. Fortunately the latest cable installations will allow up-link capability from 750Kbps to 1.5Mbps range over the same cable infrastructure. A typical cable networking infrastructure providing fast Internet access is illustrated in Figure 4-5. ATM will play a role providing robust backbone data network after the head-end where the Internet traffic is separated from the TV broadcast signals.

Cable operators will likely have to rely on other companies to provide Internet content since they do not have the necessary expertise. It would be very difficult for the cable operators to compete against Internet service provider giants such as AOL. Additionally, the cable user would prefer to have a choice in terms of selecting their own ISP. They would rather utilize the cable as a fast access medium to the Internet. In such a scenario, the cable operators will charge an access fee which will be part of the Internet bill sent by the ISPs.

One of the drawbacks of the cable network is that it is a shared medium and its bandwidth capacity is not scalable. Hence, as the number of users increase it will not be possible to provide 20 to 30 Mbps maximum access bandwidth for each user. In this case, it is most likely that cable operators will choose a pricing structure which puts a premium on guaranteed maximum access bandwidth.

V. INTEGRATION OF VARIOUS TRANSPORT TECHNOLOGIES

One of the strong features of ATM is to provide internetworking among various networks such as LAN and WAN networks, thus expanding the reach of these networks beyond their original domains. In addition, by adapting SONET/SDH as its primary physical transport medium, ATM is able to leverage vast amounts of SONET/SDH installed base to offer internetworking capability on such a grand scale.

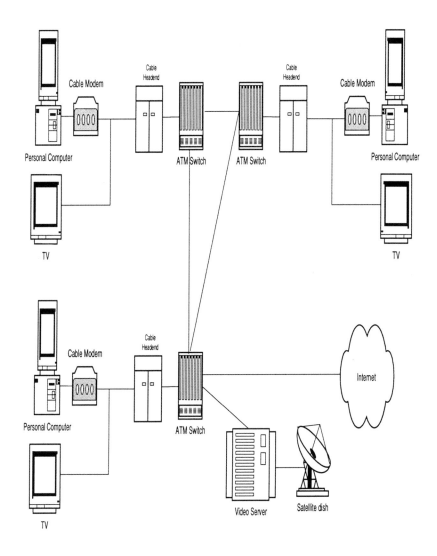

Figure 4-5 Cable-ATM inter-working for fast Internet access and video service.

Use of SONET/SDH for ATM Transport

The SONET/SDH transport technologies have been chosen to be the physical layer for ATM broadband applications by the ATM Forum and ITU-T standard organizations. The ATM protocol sits on top of the SONET/SDH-based physical layer. The key factors for choosing the SONET/SDH platform, even though SONET/SDH is a synchronous transport system, for the ATM physical layer are:

- SONET/SDH has been extensively installed around the world. Thus, ATM can take advantage of such a large installed base provided by the PSTN operators.

- It is a highly scalable transport technology due to use of fiber optics. It can handle traffic capacity from 1.5 Mbps DS1 rate to 10 Gbps OC-192 rate. As fiber optic technology advances it will be possible to provide an even higher capacity.

- It is a highly reliable transport technology.

The mapping of ATM cells into SONET/SDH payload for various speeds has been standardized. Currently, the specifications define the following SONET/SDH rates for ATM:

- DS1/E1 (1.5/2 Mbps)
- OC3c (155 Mbps)
- OC12c (622 Mbps)

LAN Internetworking

One of the key applications of ATM technology is to provide internetworking capability between various LAN technologies such as Ethernet, Fast Ethernet, and Token Ring. The LAN Emulation (LANE) protocol for ATM was specifically designed to allow interconnection among these different LAN technologies. Hence, increasing connectivity among many LANs based on different LAN technologies. Through the LANE protocol it is possible to establish large and geographically diverse Virtual LANs.

The LANE protocol also offers unprecedented bandwidth scalability through the underlying ATM network to the LAN environment. The current LAN technologies, shared or switched, offer limited bandwidth per user due to sharing and/or limitation of the physical layer transport technology being used. For example, a switched Ethernet is limited to 10 Mbps per user or Token Ring limits the maximum bandwidth to 4/16 Mbps. On the other hand, due to use of SONET/SDH as the physical layer transport, ATM can scale to 622 Mbps per user. It is expected that the 2.4 Gbps (OC-48c) rate per user will also become available very soon.

For the campus environment, the following (non SONET/SDH) ATM rates are also available:

- ATM 25 (25 Mbps)
- ATM 100 (100 Mbps)

Frame Relay-ATM Interworking for WAN Applications

Prior to recognition of ATM as the preferred network platform for WAN application, Frame Relay gained significant market penetration for WAN application. The success of Frame Relay in the WAN segment can be attributed to the following factors:

- Simplicity of the protocol.

- Very efficient use of existing PSTN networks through T1/E1 and fractional T1/E1 interfaces. Hence, low build-up cost.

- Very cost-effective data transport service compared to other alternatives such as leased lines due to sharing of physical transport links.

The Frame Relay protocol has its origin from the X.25 protocol. In order to make it suitable for fast data traffic a significant amount of overhead associated with the X.25 protocol was eliminated. For example, node-by-node error checking and flow control was delegated to end-to-end error checking and flow control instead.

At the present time, the majority of Internet traffic is being carried over Frame Relay networks. However, as the ATM technology becomes more affordable, it is expected that the Internet traffic will shift to ATM networks. The key factor for the switch from Frame Relay to ATM is due to limitation of scalability for higher bandwidths associated with Frame Relay technology.

Chapter 5

ATM PROTOCOLS

Protocols for computer communications are defined as sets of rules and message exchanges between computer systems. In order to talk to another computer, it is essential to understand its communication protocol. Typically, various communication functions involved in networking are divided into specific tasks and spread among the layers. The most well-known protocol architecture is the seven-layer Open Systems Interconnection (OSI) model.

A *protocol stack* is a set of layers that incorporates standards of one system or another at each layer. There are many different kinds of protocol stacks. The connector on the back of a computer or the communications board is referred to as the lowest layer while the applications program running in the local memory of the computer (or the network device) is referred to as the highest layer. The protocol stack bridges the gap between the hardware (i.e., the connector) and the software (i.e., the applications program).

As shown in Figure 5-1 a protocol stack includes a variety of components necessary for providing a complete solution for the user's networking problems. Signaling protocols are necessary for controlling the network. Other components include a method for monitoring the network performance and a means for managing both, the resources and the traffic on the network. It also includes a protocol for internetworking with older networking technologies.

The protocol stack can often be proprietary, i.e., wholly developed and owned by a private company, and the company is under no obligation to reveal the internal functional details to anyone. IBM's System Network Architecture (SNA) is an example of a proprietary protocol. There are other protocols that are available to everyone and referred to as an open standard. ISO and the popular Transmission Control Protocol/Internetwork Protocol (TCP/IP) belong to this category where documentation is available from a variety of sources.

I. ATM PROTOCOL STACK

ATM is an open standard based on the documentation available from the ITU-T and the ATM forum. It is an ambitious network architecture that includes support for not just the data, but other multimedia services such as voice and video. Thus it is important that signaling, management, etc. operate not only in a data networking environment but also in an equally efficient manner for all other services.

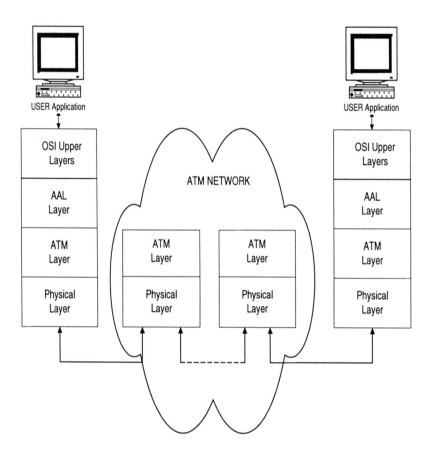

Figure 5-1   Example of ATM protocol stack application.

ATM protocol stack is for communications over a high-speed network where information (i.e., voice, video, audio and data) is sent in an unchannelized fashion through cells. Figure 5-2 shows the three primary layers of the ATM protocol reference model. The physical layer interfaces with the actual physical medium and performs the task of transmission convergence. The cell structure occurs at the ATM layer. The virtual paths and virtual channels are also defined at this layer. Support for higher layer services such as circuit emulation, frame relay, Internetwork Protocol (IP), SMDS, etc. is provided by the ATM adaptation layer.

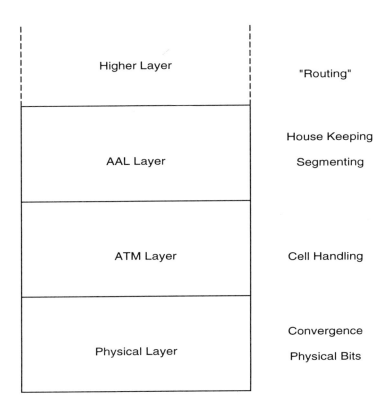

Figure 5-2 ATM layers implementing the various cell functions.

Figure 5-3 depicts the number of instances of defined standardized protocols or interfaces at each layer by boxes. It is clear that ATM is the pivotal protocol, such that for a large number of physical media, several AALs and an ever-expanding set of higher layer functions there is only one instance of the ATM layer. Thus ATM allows machines with different physical interfaces to transport data independently of the higher layer protocols using a common well-defined protocol.

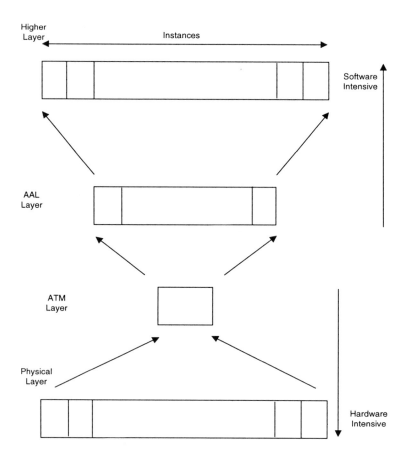

Figure 5-3 Hardware to software progression for the ATM protocol model. Number of instances in each layer is indicated by the boxes.

Figure 5-3 also illustrates how ATM implementation moves from being hardware intensive at the lower layers (the physical and ATM layers) to software intensive at the higher layers (AALs and higher layers). Table 5-1 shows the various ATM layers and their functions.

II. THE PHYSICAL LAYER

ATM is a very sophisticated protocol and technology but ultimately it deals with the transmission of digital information through a physical medium that converts two ATM devices. The physical layer is divided in two sub-layers:

1) The Physical Medium Dependent (PMD) sub-layer which provides the actual transmission of the bits, in the ATM cells.

2) The Transmission Convergence (TC) sub-layer which transforms the flow of cells into a steady stream of bits and bytes for the transmission over the physical medium.

A. PMD Sub-layer

The original transport medium for ATM was defined by ITU-T for B-ISDN architecture in the late 1980s. It was a fiber-based, scalable, high-speed network scheme referred to as Synchronous Digital Hierarchy (SDH). The SDH is standardized for use in the U.S. by the American National Science Institute (ANSI) and is known as the Synchronous Optical NETwork (SONET)

By 1992 the ATM forum declared that ATM did not have to be limited to high-speed fiber medium and other media such as coaxial cable, microwave links, etc. should be supported. Goralski (1995) and Ginsberg (1996) discuss various mediums in detail with the emphasis on the fiber-based network. The overall philosophy of ATM network is that they should be fast and highly reliable. (i.e., the transmission errors should be fewer in magnitude). But it is also important that the transport media should be readily available. ATM is meant for providing unchannelized network support to a variety of users with connections having different speeds. Thus it is more informative to group the allowed transport by physical medium rather than by technology. Figure 5-4 illustrates this framework where some transport media are meant to be used in private ATM networks while others are anticipated to be used only in public ATM networks.

B. ATM Transmission Convergence (TC) Layer

On the transmission side of the network the TC sub-layer is responsible for loading the ATM cells into the physical transport transmission frame, (i.e., SONET or DS-3). In some cases it associates cells into blocks (4B/5B, 8B/10B, etc.) before transmission. On the reception side it must delineate the individual cells in the received bit stream from the transmission frame format. It also provides error checking via the Header Error Check field in the ATM cell header.

Table 5-1 ATM layers and their functions

LAYER	SUBLAYER	FUNCTIONS
ATM Adaptation Layer (AAL)	Convergence Sublayer (CS)	• Convergence functions
	Segmentation and Reassembly (SAR)	• Segmentation and reassembly functions
ATM Layer		• Cell multiplexing and demultiplexing • Cell VPI/VCI translation • Cell header generation/extraction • Generic flow control
Physical Layer (PL)	Transmission Convergence (TC)	• Cell delineation • Cell rate coupling • HEC generation/verification • Transmission frame generation/recovery • Transmission frame adaptation
	Physical Medium (PM)	• Bit timing • Physical medium management

Following are the major functions carried out by the TC sub-layer (see Figure 5-5):

1) Transmission frame generation on the transmit side depending on the type of transport and recovery. It is also responsible for unpacking the cells at the receiver side.

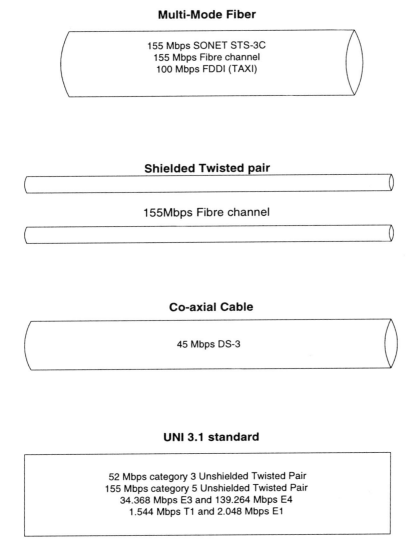

Multi-Mode Fiber

155 Mbps SONET STS-3C
155 Mbps Fibre channel
100 Mbps FDDI (TAXI)

Shielded Twisted pair

155Mbps Fibre channel

Co-axial Cable

45 Mbps DS-3

UNI 3.1 standard

52 Mbps category 3 Unshielded Twisted Pair
155 Mbps category 5 Unshielded Twisted Pair
34.368 Mbps E3 and 139.264 Mbps E4
1.544 Mbps T1 and 2.048 Mbps E1

Figure 5-4 Various forms of physical transports allowed by the ATM forum.

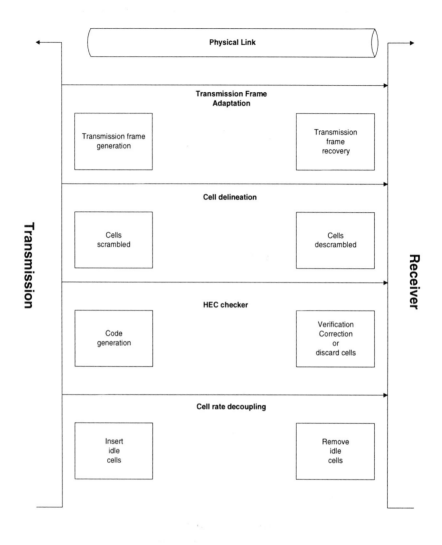

Figure 5-5 Major functions carried out by the TC sub-layer.

2) Generation of Header Error Code (HEC) sequence at the time of transmission and using it to detect and correct errors on the receiver side.

3) It is critical for the ATM layer to operate with a wide range of physical interfaces with different speeds. The TC sub-layer inserts special idle cells in order to maintain a fixed bit rate when the ATM layer is not providing cells. This is referred to as the *cell rate decoupling*.

4) Cell delineation function is necessary to make sure that receivers detect cell boundaries from a continuous stream of bits.

The cell rate decoupling feature provides the ATM network great flexibility in connection speeds over the same physical line rate. For the synchronous scheme it is essential that a fixed number of frames (bits) per second are sent over a framed physical transmission path. Consider a situation where a SONET link STS-3c (155Mbps) is connected to a customer site. However the customer (i.e., the transmitter) is generating data at a much slower rate. For example, the data bits are placed into cells at only 78 Mbps instead of the ATM network link rate of 155 Mbps. In this case, as illustrated in Figure 5-6, extra 77 Mbps of idle cells are inserted by the TC sub-layer and removed by the receiver. The idle cells have a fixed format as shown in Figure 5-7 which makes it easy for the receiver to identify and discard them.

Before accepting a cell at the receiver node it is important to check for the correct header information. This is achieved by using the HEC to supply robust error correction and detection for errors in the header (see Figure 5-8). It is important to note that error detection and correction are performed only on the ATM cell header and not on the actual data or information in the cell itself.

As illustrated in Figure 5-8 only single bit errors are corrected, which is an optimal solution in the case of fiber media where most errors are single bit. However, it may lose its effectiveness on other media, especially copper wires, with burst error characteristics.

Figure 5-9 shows the flow chart for an error correction algorithm while Figure 5-10 shows the corresponding state diagram.

III. THE ATM LAYER

The physical layer in the ATM protocol stack sends and receives cells. However, these cells are actually processed by the ATM layer. Typically, at the endpoints, the ATM layer, which deals with a stream of cells, multiplexes and demultiplexes, while at the network nodes, it switches cells.

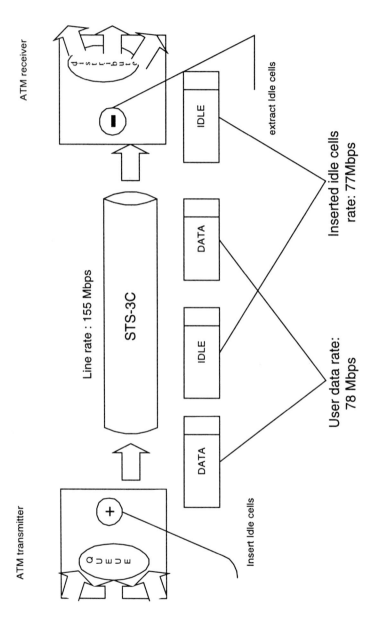

Figure 5.6 Idle cell insertion for cell rate decoupling

The ATM layer has to deal with a variety of cells, i.e., signaling cells, information (data-carrying) cells, network management cells, etc. The ATM layer examines these cells (mostly the header, but sometimes even the contents) and then decides what action must be taken. In ATM networks a switch is a network relay device that takes a cell from an input device, interprets relevant address information and gets the cell onto an output port. Unlike TCP/IP, which is a router-based connectionless protocol, ATM is a switch–based, connection–oriented protocol.

A. ATM Layer Functions

The ATM layer provides a variety of functions, such as:

1) Connection Admission Control (CAC) including connection assignment and removal
2) Cell construction, reception and header validation
3) Cell payload type discrimination
4) Cell relaying, copying and forwarding using the VPI/VCI
5) Cell multiplexing and demultiplexing using the VPI/VCI
6) Cell loss priority processing
7) Usage parameter control (UPC)
8) Support for multiple QOS classes
9) Generic flow control (GFC)
10) Feedback controls through Operation And Maintenance (OAM)
11) Explicit Forward Congestion Indication (EFCI)

	Header					Information Field
	Octet 1	Octet 2	Octet 3	Octet 4	Octet 5	
in HEX	00	00	00	01	Valid HEC	6C6C6C6C......(repeating)

Figure 5-7 Idle cell format for cell rate decoupling at the TC sub-layer.

ATM layer is where the cell headers are actually built. The ATM network nodes act based on that information. The VPI and VCIs are interpreted at the ATM layer so that the cell can be sent to the proper destination across the network. The mixing of cells for voice, video and data on the UNI also takes place at this layer which are then separated at the destination.

As discussed in Chapter 6, construction of a virtual path (VP) and virtual channel (VC) is a key concept in ATM. In general, an ATM device may be either a connecting point or an endpoint for a VP or VC. As shown in Figure 5-11, a virtual path connection (VPC) or a virtual channel connection (VCC) can exist only between two endpoints. Similarly, a VP link or a VC link can exist only between either connecting points or a connecting point and an endpoint.

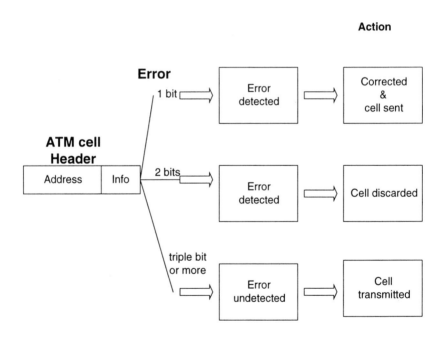

Figure 5-8 Performance of the Header Error Checker for ATM networks.

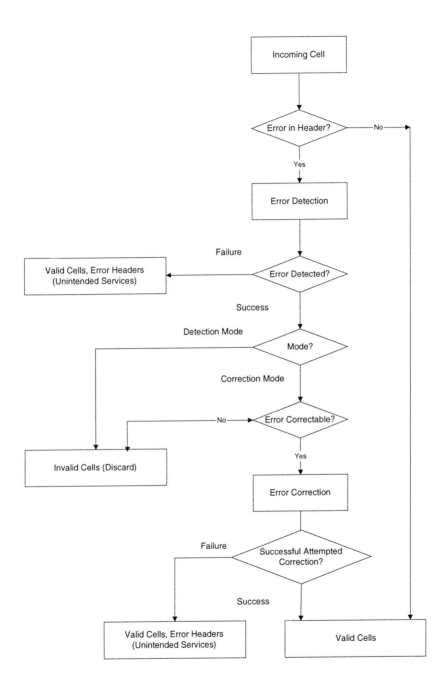

Figure 5-9 Flow chart for the Header Error Correction algorithm.

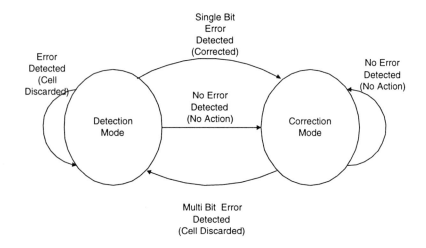

Figure 5-10 State diagram for the Header Error Check mechanism.

A VPC or VCC is an ordered list of VP or VC links, respectively. Thus a VCC defines a unidirectional flow of ATM cells from one user to one or more other users. The ATM layer requires that the network delivers the cells in the same order in which they are sent. In other words, the cell sequence integrity for a VCC must be preserved. However, the standards do not require a network to preserve the cell sequence integrity for a VPC.

The physical layer of the ATM protocol architecture has a different implementation based on different media types. Similarly the ATM adaptation layer is, generally, active and present only at the end nodes of the ATM network and does not have access to the cell header information. As a result, the ATM layer is responsible for performing the generic (general) functions, i.e.,

Network Resource Management (NRM),
Connection Admission Control (CAC),
Priority Control (PC),
Usage Parameter Control (UPC), and
Traffic Shaping (TS).

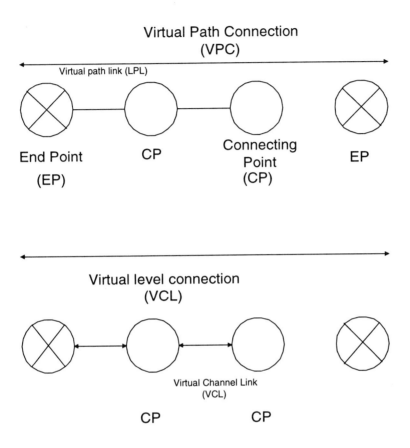

Figure 5-11 Virtual paths and Virtual channels between End Points (EP) and Connecting Points (CP).

No standard mechanisms have yet been defined to provide for these functions. These generic functions are supposed to provide the quality of service (QOS) to the users of the ATM network connections based on their need and expectations. Some of these functions are discussed in greater detail in Chapter 8, which deals with ATM network management.

Table 5-2   Traffic cell types

Cell Type	VCI	VPI
User Data	N>31	N>0
Unassigned Cell	0	0
Meta-Signaling	1	0
Remote: Meta-Signaling	1	N>0
Broadcast Signaling	2	0
Remote: Broadcast Signaling	2	N>0
PT-PT Signaling	5	0
Remote: PT-PT Signaling	5	N>0
VC OAM	3 or 4	0
SMDS Cell	15	N>=0
ILMI Cell	16	0

In the previous section on the physical layer, the function of cell-rate decoupling for framed transport was discussed. However, the ATM forum allows for other transport media for ATM cells besides framed transports. The ATM forum has recommended that cell-rate decoupling should be done at the ATM layer, using special cells called *unassigned cells.*

B.   ATM Cell Types

Certain cells have a function other than carrying user data or information across the network. These cells must be properly identified and processed appropriately while dealing with a cell stream. The operation and maintenance (OAM) cell is such a special cell, which can be identified by certain cell header value.

At the ATM layer special cells have reserved values for the VPI and VCI fields. These various combinations of VPI and VCI values (see Table 5-2) form a class known as VP/VC traffic cells. Types of cells that fall into this category are listed below:

Signaling cells
Virtual Channel OAM cells
Metasignaling cells (signaling to control other signaling connections)
SMDS cells (in case of connectionless ATM)
Interim Layer Management Interface (ILMI) cells, etc.

Figures 5-12 and 5-13 illustrate the special cells with *predefined header field values*, which exist in the physical layer. Note that the last few bits of octet 4 are reserved for use by the ATM physical layer.

IV   ATM Adaptation Layer (AAL)

ATM networks are designed for providing transport services to applications which use a wide variety of formats, and information types (Data, Video, Voice, etc.). The majority of these applications and their related protocols generate data units of variable sizes before sending the applications data across the ATM network as they have to be *adapted* to the ATM network. At the transmitter the variable-length data must be segmented into cells. At the receiver these cells must be reassembled back into original variable length data. This entire preparation is carried out by the ATM Adaptation Layer (AAL) which is the highest level of the ATM protocol stack.

A. AAL Structure

The structure of the AAL and its interfaces are illustrated in Figure 5-14. The various sub-layers within the AAL are shown along with the primitives regarding their respective protocol Data Units (PDUs) that are passed between them. The AAL is divided into the segmentation and Reassembly (SAR) sub-layer and the convergence sub-layer (CS). The ATM cell payload, also called the SAR-PDU primitives, is passed to and from the ATM layer through the ATM Service Access Point (ATM-SAP).

The CS is further divided into Common Part (CP) and service specific (SS) components. The primitives regarding the AAL PDUs are passed through the AAL-SAP (shown in Figure 5-14 at the top) to the higher layers.

The SSCS is responsible for a series of actions that must be taken for providing the services demanded by the quality of services (QOS) parameters. In some cases the SSCS is further divided into Service Specific Coordination Function (SSCF) and Service Specific Connection Oriented Protocol (SSCOP) components.

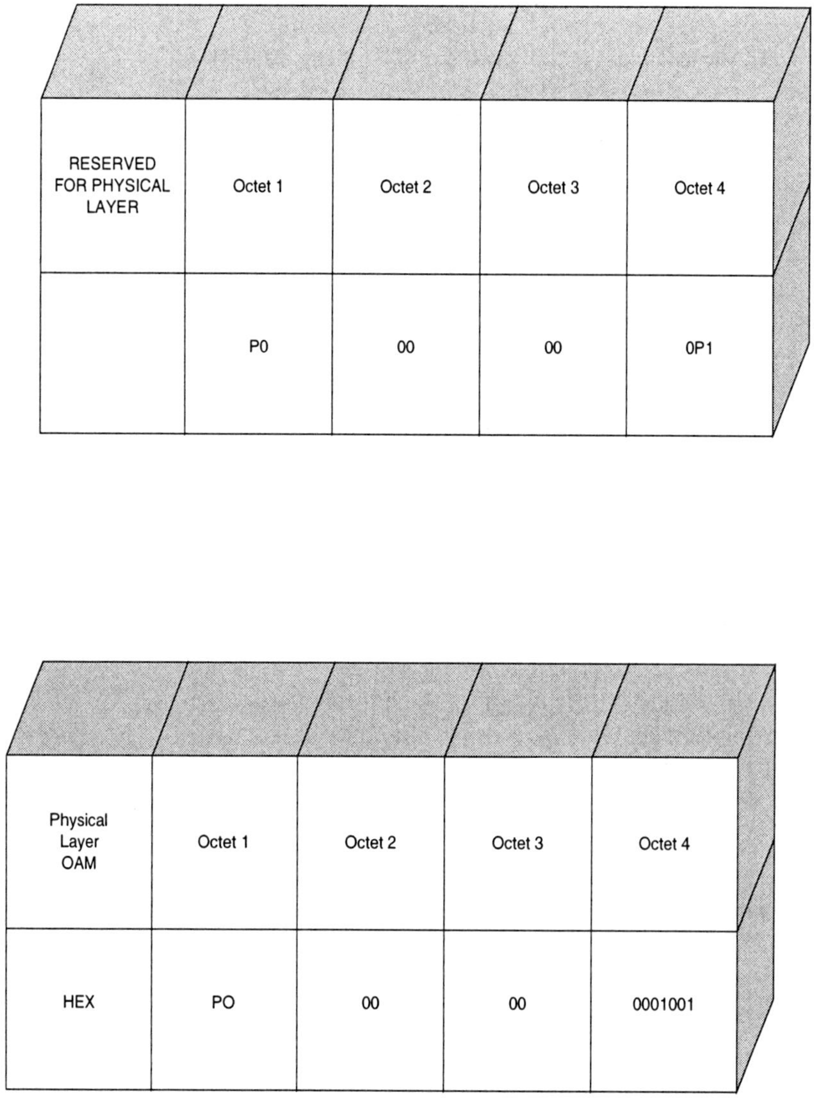

RESERVED FOR PHYSICAL LAYER	Octet 1	Octet 2	Octet 3	Octet 4
	P0	00	00	0P1

Physical Layer OAM	Octet 1	Octet 2	Octet 3	Octet 4
HEX	PO	00	00	0001001

Figure 5-12 Special cells.

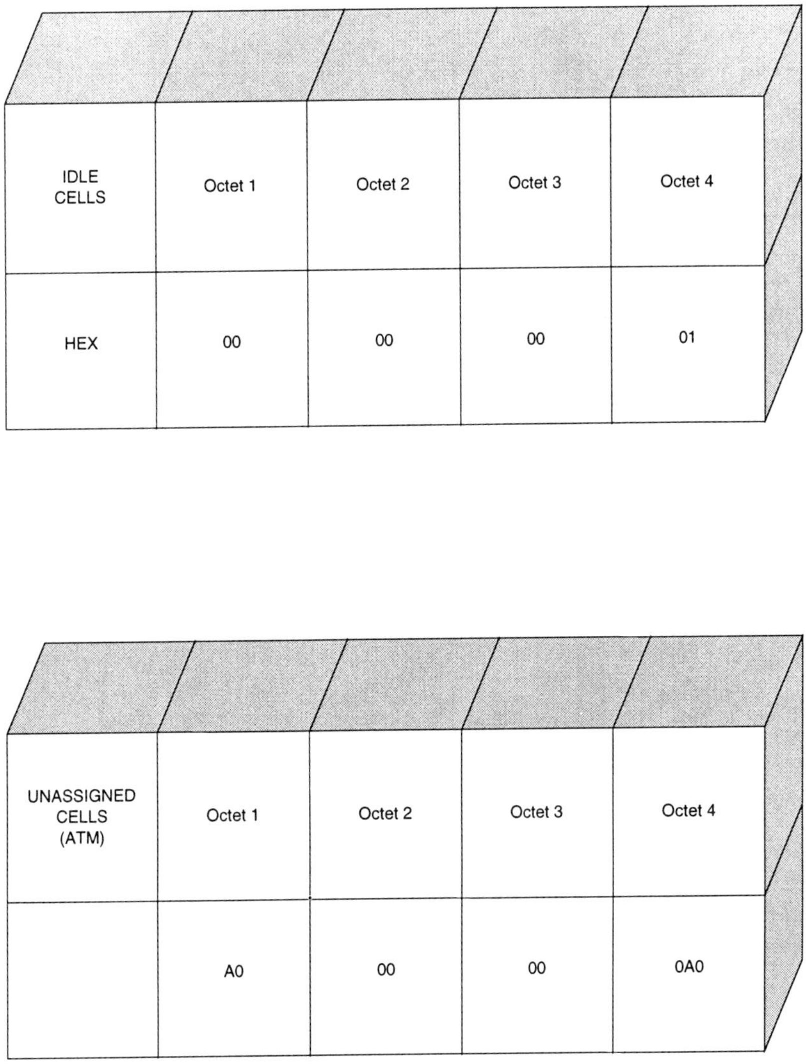

IDLE CELLS	Octet 1	Octet 2	Octet 3	Octet 4
HEX	00	00	00	01

UNASSIGNED CELLS (ATM)	Octet 1	Octet 2	Octet 3	Octet 4
	A0	00	00	0A0

Figure 5-13 Some more special cells.

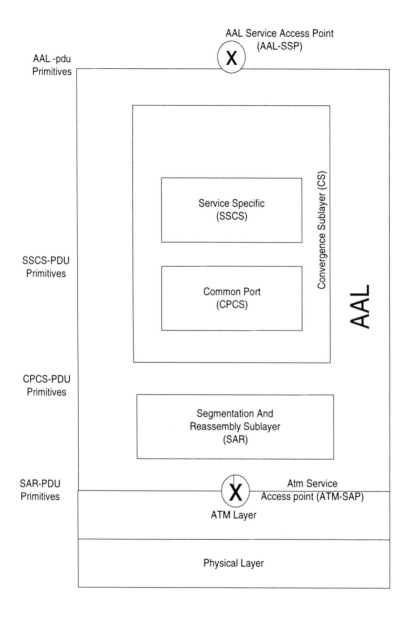

Figure 5-14 A generic model for the AAL protocol sub-layer.

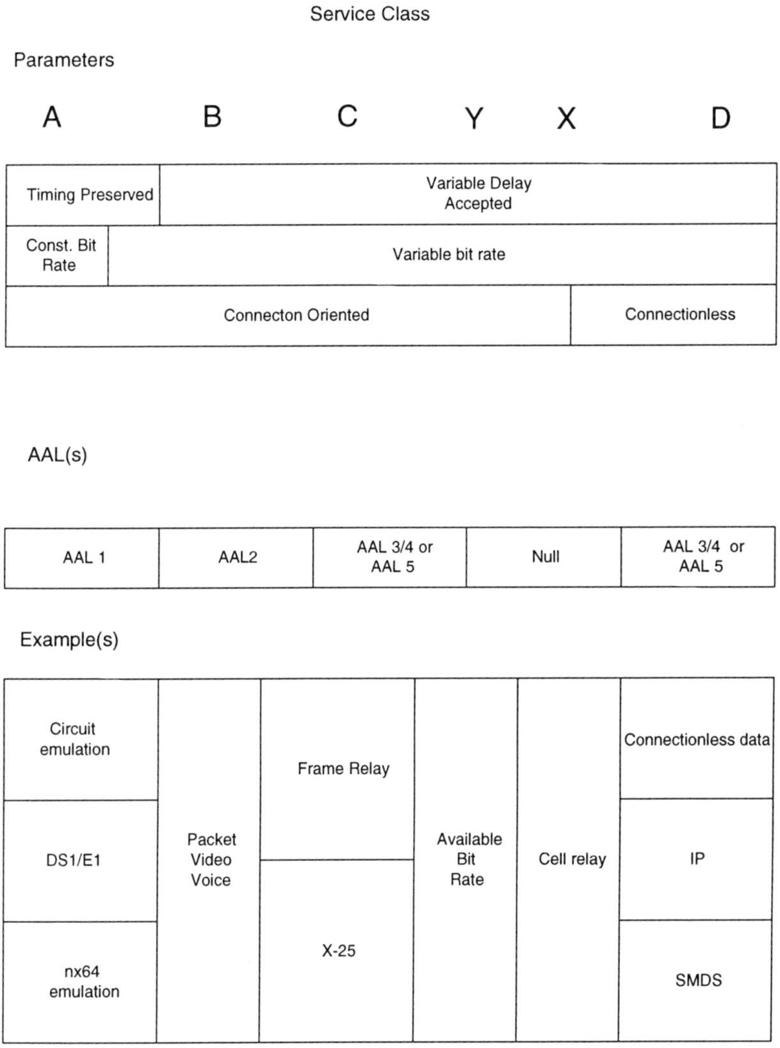

Service Class

Parameters

A	B	C	Y	X	D

Timing Preserved	Variable Delay Accepted				
Const. Bit Rate	Variable bit rate				
Connecton Oriented				Connectionless	

AAL(s)

AAL 1	AAL2	AAL 3/4 or AAL 5	Null	AAL 3/4 or AAL 5

Example(s)

Circuit emulation	Packet Video Voice	Frame Relay	Available Bit Rate	Cell relay	Connectionless data
DS1/E1					IP
nx64 emulation		X-25			SMDS

Figure 5-15 ATM service classes.

B. Classification Based on AAL Service Attributes

ITU-T originally defined the basic principles and classification of AAL functions. The ATM Forum has proposed additional service classes. The classification is based on the following attributes:

1. Timing relationship between source and destination
2. Bit rate
3. Connection Mode

Figure 5-15 illustrates the four ITU-T defined service classes labeled A through D. Class A is constant bit rate (CBR) while C through D are variable bit rate (VBR). Classes X and Y are the additional two proposed by the ATM forum. In class X cells are transported as they are presented to the ATM layer; thus it is defined as an Unassigned Bit Rate (UBR) Service. In the case of class Y the cells are transported across the ATM network if there is capacity available for them. As a result, class Y is defined as Available Bit Rate (ABR) Service.

The various service classes shown in Figure 5-15 result in six types of services. The null type of AAL (AAL-0) is used for cell relay services that are inherently cell based and need no adaptation. The rest are labeled AAL-1, AAL-2, and so forth.

C. ATM Adaptation Layer Functions

ATM cells consist of 5 byte headers and 48 byte payloads. ATM cell headers are used for multiplexing and demultiplexing as well as switching based on the connection number at the ATM layer. The AAL only deals with payloads so there are no cells at this layer. The cell header (as discussed in Section II of this chapter) is generated at the transmitter to the ATM layer and stripped at the receiver.

As seen in Figure 5-15 ATM supports various service classes. Thus AAL is composed of service classes. Thus AAL is composed of a series of compartments and it is modular in structure. Each compartment contains a service dependent function and some other functions that are common to all ATM network applications. At the transmitter, based on the required user service the corresponding active AAL passes on the payloads to the ATM layer. At the receiver the ATM layer is responsible for demultiplexing the cells based on the connection number (VPI/VCI) and passing on the payloads to the proper AALs.

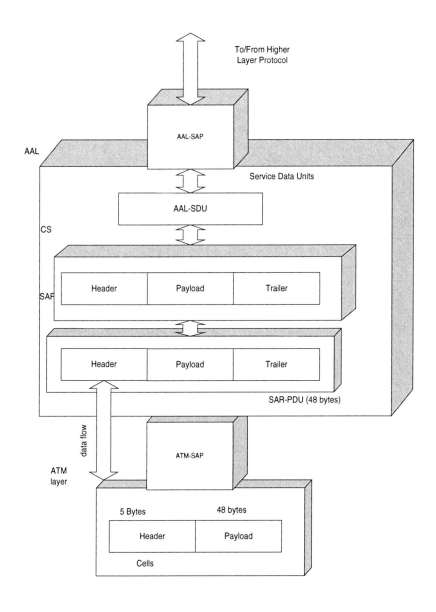

Figure 5-16 General AAL functions.

At the transmitter the user data flowing into the ATM network through the AAL are often in variable length data units. Thus the data arriving from the SAR sub-layer look pretty much like what come in from the higher layer. At the SAR sub-layer the variable length data units are segmented into a series of 48 byte payloads.

At the receiver again SAR is responsible for reassembling these sequences of 48 byte payloads, transmitted across the ATM network into variable length data units. Thus the AAL shields the user application from having to worry about the inner working details of the ATM network.

The convergence sublayer (CS) interacts with the higher layer protocols (or user application) through the service access point (SAP). It sets up the data stream based on the service requirements of the application and sends it on to the SAR for segmentation.

As illustrated in Figure 5-16 variable length higher layer Protocol Data Units (PDUs) enter the AAL (i.e., the ATM Protocol stack) via the service access point (SAP) Here it is converted to an AAL service data unit (SDU) by adding application specific headers and trailers. Then it is passed in the SAR where it is broken up into 48 byte SAR-PDUs They include some additional headers and trailers that are placed on these fragments. The data are then passed on to the ATM through the ATM-SAP and cells are composed by adding 5-byte cell headers.

Chapter 6

BANDWIDTH ALLOCATION IN ATM NETWORKS

Flexible and reliable bandwidth allocation is one of the key features of ATM. In terms of bandwidth allocation capability, ATM represents a somewhat more rational approach with respect to both ends of the spectrum. At one end of spectrum is the very rigid synchronous SONET/SDH technology. At the other end of the spectrum is shared medium Ethernet technology which allows bandwidth usage with no restriction at all.

SONET/SDH has been initially designed to multiplex voice circuits. The smallest unit of bandwidth allocation possible is at the T1 rate (1.5 Mbps) for SONET and E1 rate (2 Mbps) for SDH. Without any slience suppression, voice samples flow through the SONET/SDH network at a constant rate, i.e., 64 Kbps per voice call. However, when it comes to carrying bursty data traffic, a significant amount of the allocated bandwidth for a particular connection is wasted. In addition, the bandwidth has to be allocated according to peak rate. Hence, if the peak rate for a data connection is 3 Mbps, then the next available payload greater than 3 Mbps is the DS3 rate (45 Mbps) for SONET. In this case we would be wasting minimum 42 Mps bandwidth capacity.

Ethernet technology allows a maximum level of sharing of available bandwidth on a physical medium. Hence, on a typical 10 Mbps Ethernet bus, each user attached to the bus is capable of using the total 10 Mbps bandwidth. However, due to lack of coordination, conflicts arise among the users as a result of simultaneous attempts to access the bus by multiple users. Therefore, during a heavy traffic period, total available bandwidth is hardly accessible to any of the users.

ATM technology, on the other hand, partitions the physical medium bandwidth logically and creates sub-rates at a very fine granularity. Each user is assigned a logical bandwidth pipe according to its needs. Through this logical partitioning, a user is isolated from the other users. Hence, ATM allows guarantee of service at the negotiated rates to every user. We will discuss in greater detail how ATM manages bandwidth allocation, types of bandwidth allocation strategies and bandwidth allocation classes later in this chapter.

I. VARIABLE BANDWIDTH ALLOCATION

Variable bandwidth allocation is one of the cornerstones of ATM technology. Through the use of Virtual Path (VP) and Virtual Channel (VC) concepts, ATM is capable of logically partitioning the available bandwidth on a physical medium such as fiber optics, copper and coaxial cables, and ATM switches. Each of these VPs and VCs represents a logical pipe which carries user traffic. Typically, during the path setup, each user is allocated a VP or VC based on the negotiated bandwidth as part of the traffic contract. Through the use of ATM traffic management, the terms of these traffic contracts are enforced. For example, the ATM traffic management ensures that users do not exceed the limits of their bandwidth allocations.

The VP and VC mechanisms allow allocation of bandwidth to each user at a very fine granularity so that the available bandwidth of the physical medium is utilized very efficiently. As mentioned earlier, one of the main drawbacks of synchronous SONET/SDH technology is the fact that it allocates bandwidth in fixed quantities called Virtual Containers. Thus, if the user request does not exactly match the bandwidth size of the allocated Virtual Container, then the unused portion of the allocated bandwidth is wasted. Fortunately, by using ATM on top of the SONET/SDH transport, efficient usage of the physical bandwidth capacity can be achieved.

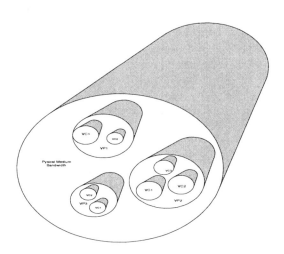

Figure 6-1 Flexible bandwidth allocation via VP/VC mechanism.

II. VIRTUAL PATH AND VIRTUAL CHANNEL CONCEPTS

The VP/VC concept of ATM allows creation of logical pipes on a single physical transport medium. Although users are sharing the same physical medium, the VP/VC concept limits the amount of traffic each user can generate according to the negotiated traffic contract between the user and the ATM network. The ATM cells from the users who exceed their traffic contract are either immediately discarded at the control point or marked as a candidate for discard by downstream control points in case of network congestion.

Unlike the shared medium-based Ethernet technology, ATM provides a certain degree of control via VP/VC mechanism for the amount of traffic each user can generate on the shared medium. Hence, ATM is capable of providing a quality of service (QoS) guarantee, a very desirable feature which the Ethernet technology can not provide in its present form. This is one of the most significant distinctions between ATM and LAN technologies such as Ethernet. This distinction becomes more apparent when dealing with delay sensitive real-time voice/video traffic. However, today, there are attempts being made to provide similar QoS guarantees for Ethernet by using a reservation scheme called RSVP.

ATM defines the set of traffic classes and traffic parameters to characterize incoming traffic. The traffic classes include Constant Bit Rate (CBR), Variable Bit Rate (VBR), Available Bit Rate (ABR) and Unspecified Bit Rate (UBR). The traffic parameters include Peak Cell Rate (PCR), Sustained Cell Rate (SCR), Minimum Cell Rate (MCR), and Maximum Burst Size (MBS). These traffic classes and parameters are used to define a traffic contract between the ATM network and the user. Through monitoring these parameters, ATM network enforces traffic contracts. These concepts are described in ATM Traffic Management specifications. We discuss these concepts in more detail in Chapter 8.

We next look at the traffic classes offered by ATM to handle different bandwidth demands.

III. CONSTANT BIT RATE

The Constant Bit Rate (CBR) service class is used for delay-sensitive, real-time traffic such as voice and video. It is also used for carrying traditional synchronous transport such as T1 traffic over ATM networks. This is called Circuit Emulation Service (CES). The AAL1 ATM adaptation layer handles the circuit emulation function.

It is necessary to use the CBR service class to guarantee certain delay, delay variance and jitter limits associated with the synchronous traffic. Only one parameter, PCR, is used to describe and monitor user traffic for the CBR class. If the user traffic exceeds the PCR limit, then the user cells are marked for possible discard by setting the CLP parameter in the ATM cell header to 1. This behavior is illustrated in Figure 6-2.

The CBR service class allocates bandwidth according to the peak rate of incoming traffic. The user is guaranteed full access to the allocated peak bandwidth throughout the established connection and is not influenced by the bandwidth usage patterns of other users. Hence, it is possible to carry both synchronous traffic in the circuit emulation mode and very bursty LAN traffic on the same physical medium. This is one of the most desirable characteristics of ATM.

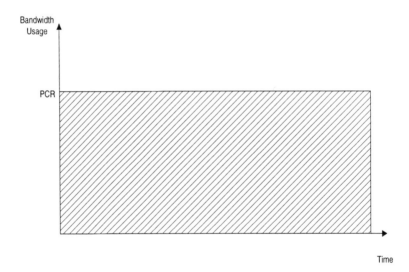

Figure 6-2 Monitoring of CBR traffic: cells above the PCR line are marked for possible discard.

IV. VARIABLE BIT RATE

The variable Bit Rate (VBR) service class primarily is used for handling bursty with varying bandwidth usage over time. For VBR, two subclasses are defined: Real Time VBR (rt-VBR) and Non-Real-Time VBR (nrt-VBR). The rt-VBR is used if the traffic source is operating in real-time and it is delay sensitive such as compressed voice and MPEG video traffic. For example, a voice application with silence suppression and voice compression capabilities can use rt-VBR instead of CBR to reduce its bandwidth usage while still maintaining the delay requirement. If the traffic source is not a real-time application, then the nrt-VBR service class is used. The difference is that the real-time traffic is less tolerant to delay than the non-real-time traffic. For example, for video conferencing rt-VBR is more appropriate. On the other hand, nrt-VBR service is more suitable when the application is not delay bound. The nrt-VBR class traffic can tolerate delay through buffering. The nrt-VBR can be used for data transfer which is not delay sensitive but requires low CLR.

The VBR service class requires several traffic parameters to describe the incoming traffic. These parameters include PCR, SBR, MBS. In addition, rt-VBR is delay bound with the max CTD parameter. The bandwidth characteristics of VBR class are shown in Figure 6-3. As long as the VBR traffic stays within the given bandwidth limits (bandwidth profile), it is not subjected to cell drop by the ATM network, i.e., the assigned CLR level is maintained by the network.

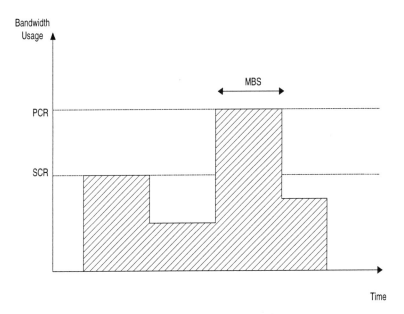

Figure 6-3 Monitoring of VBR: cells above the PCR line or exceeding MBS limit above SCR line are marked for possible discard.

V. UNSPECIFIED BIT RATE

The Unspecified Bit Rate (UBR) service category was primarily defined for handling LAN traffic to take advantage of the remaining available bandwidth. UBR does not guarantee any level of bandwidth. It allows a source to use max remaining available bandwidth. Therefore, during a low traffic period, a LAN source can take advantage of the full line speed. However, during very high bandwidth usage, it is possible that the UBR source will not receive any bandwidth at all. For UBR traffic class, ATM network does not guarantee any of the QoS parameters (CLR, CTD). The UBR service is, in a sense, considered as the Best Effort attempt by the ATM network. As one can realize, this behavior is very similar to the Ethernet LAN environment. In fact, UBR was defined to match this type of traffic pattern. In addition to LAN traffic, UBR can also be used to take advantage of the available background bandwidth for applications such as transferring news and weather pictures, file transfer and email. ATM does not consider congestion control for the UBR traffic. It is assumed that it is done by a higher layer on a end-to-end basis.

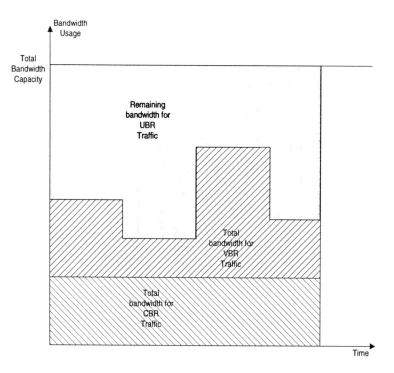

Figure 6-4 Bandwidth allocation for UBR traffic.

VI. AVAILABLE BIT RATE

The Available Bit Rate (ABR) service class has been defined as a refinement to the UBR class. It provides some degree of control over the traffic flow as well as guaranteeing a minimum bandwidth. Through the traffic flow control mechanism, the traffic source adjusts its traffic generation according to the network condition. ABR takes advantage of all the remaining bandwidth during normal traffic conditions up to the specified PCR level. The PCR parameter is used to make sure that the remaining bandwidth is shared fairly among ABR type users. However, when the congestion occurs, the ABR source is expected to reduce its bandwidth usage. Unlike UBR, ABR is guaranteed a minimum bandwidth specified by the traffic parameter Minimum Cell Rate (MCR). The ABR class is typically used for LAN traffic.

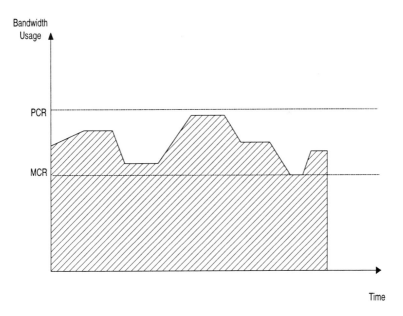

Figure 6-5 Bandwidth allocation according to ABR service class.

Chapter 7

SWITCHING ARCHITECTURES FOR ATM NETWORKS

An ideal NxN switching architecture for ATM networks can be modeled as a box with N input ports and N output ports, providing connections between its N input ports and N output ports. Figure 7-1 illustrates such an ideal switching model. The internal structure of a switching architecture determines how the connections are made between its input-output pairs. Depending on its internal structure, a switching architecture can be classified into the following categories:

- blocking or non-blocking
- self-routing or central routing
- input buffering, output buffering or central buffering
- buffer management: complete sharing, partial sharing, and complete partitioning

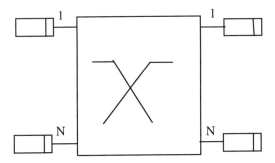

Figure 7-1 An ideal ATM switch model.

I. BUFFERING MODELS FOR ATM SWITCHING NETWORKS

Buffer management is one of the key elements of ATM cell switching and has a direct influence on switching performance. Due to statistical multiplexing used in ATM switching, an ATM switch should be capable of dealing with a wide fluctuation of incoming cell traffic. This is mainly achieved through buffering of cells which cannot be transmitted in the same time slot.

Buffering allows smoothing of peak cell traffic through storage of cells to be transmitted at a later time. Buffering in an ATM switch can take one of three basic forms: input buffering, output buffering and central buffering [Karol 1987, Karol 1988, Kim 1990, Mun 1994, Re 1993]. Input buffering stores incoming cells at the input ports of an ATM switch while output buffering stores cells at the output ports of an ATM switch. In case of central buffering, a central buffer space is shared by both input and output ports.

A. Input Buffering

With input buffering, the switching network can run at the same speed as the input/output port transmission speed. However, the performance of a switching network with input buffering and First-In First-Out (FIFO) server strategy suffers considerably due to Head-Of-Line (HOL) blocking. The maximum throughput attainable with input buffering and FIFO strategy is limited to 0.586 as N becomes large [Karol 1988].

The performance of input buffering can be improved significantly by changing the server strategy. If we allow the routing mechanism to choose cells for transmission to output ports from each input buffer within a window of W cells, then it becomes possible to choose a maximum number of cells with disjoint output destinations to be transmitted in a single time slot. In [Karol 1988], it has been shown that with such a modification, it is possible to increase the performance of input buffering to 0.88. However, the routing for such a scheme is a NP-complete problem and the solution has to be obtained within one time slot so that while currently selected cells are being transmitted the routing mechanism can select a new set of cells to be transmitted in the next time slot.

B. Output Buffering

Output buffering is superior to input buffering in terms of performance. However, output buffering requires that the switching network has to run N times faster than the input/output port transmission speed in order to be able to handle N packets arriving at the same output port within a single time slot.

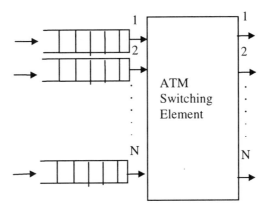

Figure 7-2 Input buffering.

C. Central Buffering

Central buffering uses a shared memory inside the switch and combines the individual output queues of the output queuing into a central queue. The behavior of central buffering is very similar to output buffering.

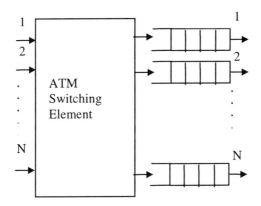

Figure 7-3 Output buffering.

111

The mean waiting time is the same as the output buffering. However, central buffering offers a better utilization of the available buffer space due to complete sharing. Thus, it requires a smaller buffer size than the buffer size required for output buffering for the same cell loss ratio. However, a more complicated control logic is necessary to maintain the order of cells when delivering cells to output ports. As in the case of output buffering, central buffering also requires the switch speed to be N times faster than the input/output port transmission speed which significantly increases the cost of the switch.

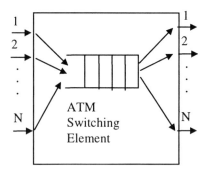

Figure 7-4 Central buffering.

II. ROUTING MODELS FOR ATM SWITCHING NETWORKS

The switching networks can be classified into two categories, blocking and non-blocking, in terms of their internal blocking characteristics. An NxN non-blocking switching network allows N input-output pairs to be connected simultaneously, given that these pairs are disjoint. The best known non-blocking switch is the crossbar switch. A typical 4x4 crossbar switch is shown in Figure 7-5. It consists of an array of 4x4 cross-point switches representing each input-output pair.

The throughput of the crossbar switch approaches to unity under ideal conditions, in which 4 disjoint routing requests are presented simultaneously to the switch at every connection interval. Because of its ideal throughput characteristics, a crossbar switch is used as a reference model for a performance study of other switching networks. In practice, due to the high cost associated with very large number of cross-points, crossbar switching networks are not suitable for a very large network structures. Therefore, alternative switching network models with an acceptable degree of blocking probability were introduced.

The output blocking in ATM switching networks is unavoidable due to the arrival of cells at different inputs destined for the same output port in the same time slot even if the switching network is internally non-blocking such as the crossbar switch. Additionally, the blocking switching networks such as the Banyan network also suffer from internal path conflicts. As a result, both output and internal blocking require buffering of cells in ATM switching networks. Buffering of packets which cannot be transferred to an output port during a current time slot can be done either at an input port or an output port as discussed in Section 7.1.

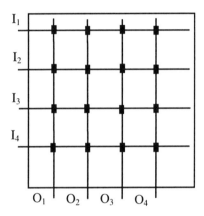

I: Input port
O: Output port

Figure 7-5 A 4x4 crossbar switch.

The routing operation in switching networks can be controlled centrally or distributed over the entire switching network, i.e., each cross-point in the network participates in the routing operation as in the case of self-routing. Typically, the central routing control is employed in the crossbar switching network whereas the Banyan network is inherently self-routing by design. The central routing requires a very fast routing control logic which is capable of setting N connections simultaneously per time slot. Self-routing due to its distributed nature is required to set up a single connection at each cross-point per time slot. Its routing speed is thus determined by the connection request arrival rate per input port. Therefore, self-routing is highly desirable for high-speed switching networks.

III. MULTISTAGE INTERCONNECTION NETWORKS (MINS) FOR ATM SWITCHING NETWORKS

One class of switching networks is multistage interconnection networks. A typical multistage interconnection network consists of a large number of elementary NxN cross-point switches ($N \leq 32$) connected through external links in a certain fashion to provide a larger switching network structure.

A. Nonblocking MINs

A large non-blocking MIN can be constructed using nonblocking switching elements such as crossbar switches. For example, a 16x16 nonblocking MIN can be constructed using 8 4x4 crossbar switching elements in a two-stage formation as shown in Figure 7-6. This particular configuration provides a single internal path for each input/output port pair. It is also possible to provide more than one concurrent internal path for each input/output port pair using additional stages and different switch sizes. For example, the number of concurrent internal paths for each input/output pair can be increased from one to four using a 3-stage configuration as shown in Figure 7-7.

B. Blocking MINs

One of the best known switching network models in this class is the Banyan network in which the elementary switch is a 2x2 crossbar switch. Figure 7-8 illustrates an 8x8 Banyan switch which is constructed from these 2x2 crossbar switches. For the 2x2 switch only 1 bit address is required. For the 8x8 Banyan switch, 3 bit addressing is required to route a cell through the switch. Each input and output port of the 8x8 Banyan switch is identified by a 3-bit address.

The cell at input port 100 shows the data field as well as the destination address field which is 010 in this case. In order to route this cell from input port 100 to output port 010, one bit is used for each stage. There is one unique path for each input-output pair. The bold line indicates the path through which the cell is routed through the switch. In the worst case situation, in an NxN Banyan switching network, during a single connection interval, $\sqrt{N}$ out of N disjoint input-output routing requests cannot be serviced due to contentions at the connection links between stages.

114

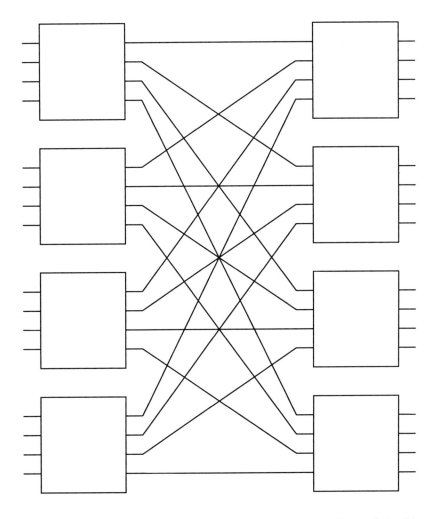

Figure 7-6 Two-stage MIN configuration for a 16x16 nonblocking switch with a single internal path for each input/output pair.

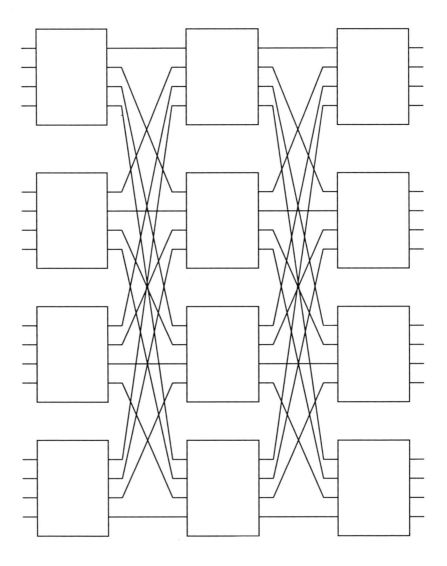

Figure 7-7 Three-stage MIN configuration for a 16x16 nonblocking switch with four concurrent internal paths for each input/output pair.

116

IV. BUFFER MANAGEMENT

Kamoun and Kleinrock [Kamoun 1980] studied the performance of five limited buffer management schemes under general traffic conditions. These five buffer management schemes can actually be reduced to three main categories: Complete Sharing (CS), Complete Partitioning (CP), and Partial Sharing (PS). The three schemes are mentioned in [Kamoun 1980]. Sharing with Maximum Queue Lengths (SMXQ), Sharing with a Minimum Allocation (SMA), and Sharing with a Maximum Queue and Minimum Allocation (SMQMA), are actually variations of the PS scheme with a varying degree of upper and lower bounds on sharing.

Both CP and PS schemes impose a hard limit on buffer usage. CP and CS are the two extremes and PS is a compromise between CP and CS. CP simply partitions the entire buffer space among users, i.e., no sharing is allowed, whereas CS allows all of the available buffer space to be shared by every user.

As indicated in [Kamoun 1980], CS scheme provides a better performance under low to medium input traffic loads and for fairly balanced traffic distribution across all input ports over CP. However, for highly unbalanced traffic patterns and for high input traffic load even with fairly balanced distribution CP and PS perform better than CS.

With respect to buffer utilization, the CS allows full utilization whereas CP and PS do not allow full utilization due to pre-allocation of buffers to certain input ports (full utilization may be possible for these schemes only in very extreme overload conditions). The buffer utilization is a significant factor in determining the required buffer size for a system under consideration. The CS scheme requires smaller buffer size than the other two schemes.

V. BROADCAST AND MULTICAST REQUIREMENTS FOR ATM SWITCHING NETWORKS

Broadcast and multicast capabilities are two essential requirements for ATM switching networks due to the nature of traffic carried by these networks. The difference between the broadcast and multicast is that broadcast multiplies an ATM cell to all destinations whereas multicast multiplies an ATM cell to only a specified number of destinations.

Two specific applications, which require broadcast/multicast capabilities, are Video/Audio broadcast and conferencing, and LAN emulation. Although broadcast/multicast of ATM cells can be achieved by the source by generating the necessary number of copies of ATM cells to be broadcasted or multicasted at the source node, this is not a very efficient approach.

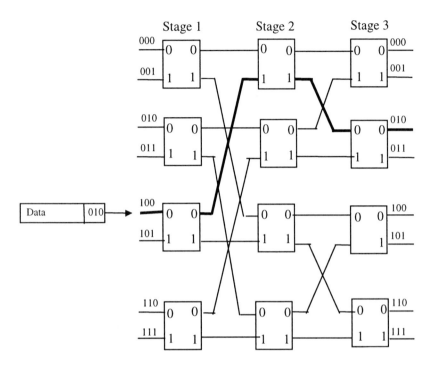

Figure 7-8 An 8x8 Banyan Switch

This approach would simply generate unnecessary traffic in ATM switching networks and possibly cause performance degradation. It is much more desirable to push the copying process as close as possible to the destination nodes in the network. Therefore, it is essential that ATM switching networks provide these capabilities at the switching hardware.

VI. ATM SWITCH ARCHITECTURES

In this section we survey the most widely used switching architectures in the industry. In general an ATM switching system is built around a switching network of varying complexity or a common ATM cell bus structure as shown in Figure 7-9.

118

Smaller ATM switches designed for workgroup, campus, and access multiplexer/concentrator applications use a small crossbar switch or ATM cell bus architecture to provide a cell path between various application specific Line Interface Modules (LIM). This approach is mainly driven by a lower cost factor. On the other hand, larger ATM switches designed for core networks use more sophisticated switching fabrics. In this case, the scalability is the main factor for choosing a particular switching fabric.

Each application segment has a particular range of bandwidth requirements. When designing an ATM system for a particular segment, the bandwidth requirement plays an important role in determining the switching architecture. Table 7-1 lists typical bandwidth requirements for some of these application segments.

A typical switching fabric is designed by combining several switching ICs in a particular formation. These switching ICs are the basic building blocks of any switching fabric. Some of these ICs require additional ICs to provide a common ATM interface to application specific LIMs.

The most common interface is the standard UTOPIA interface. As an example, we look at Siemens ATM chipsets to build various switching fabric sizes with a standard UTOPIA interface. Figure 7-10 shows a basic building block of a switching network using the Siemens chipset. It is possible to place the ASP chips on the LIM cards and have a SLIF interface between the LIM cards and the switching fabric. It is also possible to place the ASP chips on the switching modules to have a UTOPIA interface between LIMs and the switching fabric. This decision depends strictly on requirements of a particular design of an ATM system.

The core component of the Siemens switching chipset is the 32x16 ATM Switching Matrix (ASM). The block diagram of the ASM chip is shown in Figure 7-11. The ASM chip is a self-routing and nonblocking switch. It also provides a multicast function. The ASM chip uses the central buffering concept. It interfaces with other support chips via a custom ATM interface called Switch Link Interface (SLIF).

SLIF expands the standard 53 octet ATM cells into 64 octet cells by adding additional header information, synchronization, checksum and cell sequence number octets. The main purpose of the additional header information at the SLIF interface is to speed up the routing function in the ASM chip.

119

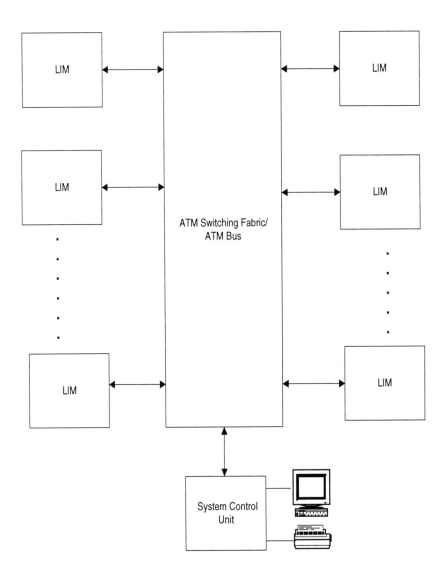

Figure 7-9 Typical ATM switching architecture.

Table 7-1 Bandwidth requirements for various application segments

Application Segment	Bandwidth Requirement
Workgroup	1.2 Gbps – 5 Gbps
Campus Backbone	2.5 Gbps – 10 Gbps
Enterprise	2.5 Gbps – 20 Gbps
WAN Backbone	2.5 Gbps – 40 Gbps
Access Multiplexer/Concentrator	155 Mbps – 2.5 Gbps
Multiservice Access Switch	10 Gbps – 40 Gbps
Core Network	10 Gbps – 1 Tbps
Backbone	> 1 Tbps

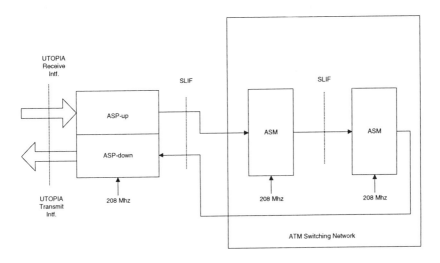

Figure 7-10 Combination of Siemens ASP-up, ASP-down and ASM chips that forms a switching unit with Utopia interface.

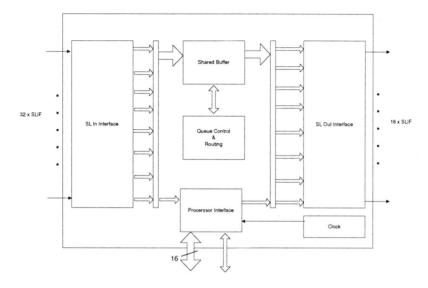

Figure 7-11 Siemens ASM chip block diagram.

The other two members of the Siemens ATM switching chipset are the ATM Switching Preprocessor Upstream (ASP-up) and ATM Switching Preprocessor Downstream (ASP-down). These two chips perform complementary functions. The ASP-up chip, as shown in Figure 7-12, provides up to four standard UTOPIA interfaces for the incoming ATM ports. Additionally, it performs the following functions:

- Format conversion from standard 53 octet ATM cell format to 64 octet SLIF format

- Traffic policing

- OAM functions

- Address translation

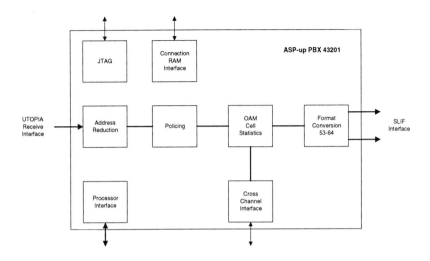

Figure 7-12 Siemens ASP-up chip block diagram.

The ASP-down chip, as shown in Figure 7-13, provides up to four UTOPIA interfaces for outgoing ATM ports. In addition, it performs the following function on the outgoing side:

- Format conversion from 64 octet SLIF format to standard 53 octet ATM cell format

- Address translation

- OAM functions

- Output buffering

- Support for multicasting

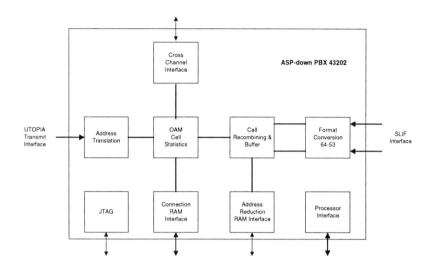

Figure 7-13 Siemens ASP-down chip block diagram.

Using the ASM chip, it is possible to design a non-blocking switching fabric to support up to a 32x32 switch dimension in a single-stage configuration. Figure 7-14 illustrates a single-stage 32x32 switching fabric using two ASM chips. Using multiple stages, the size of the switching matrix can be expanded further. For example, as shown in Figure 7-15, it is possible to construct a 64x64 switching fabric in a two-stage configuration.

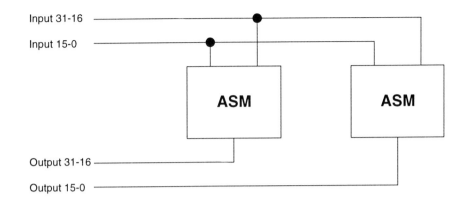

Figure 7-14 Forming a 32x32 switching fabric with Siemens ASM chips.

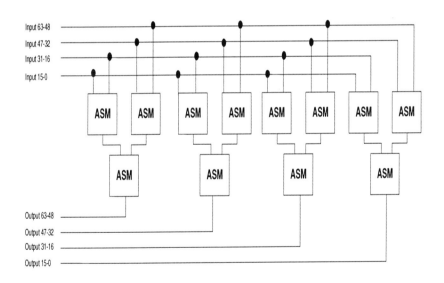

Figure 7-15 Forming a 64x64 switching fabric with Siemens ASM chips.

Chapter 8

ATM TRAFFIC MANAGEMENT: TRAFFIC ENFORCEMENT AND TRAFFIC SHAPING

It is possible that the traffic from user nodes in the ATM network can exceed its capacity, causing memory buffers of the ATM switches to overflow, leading to data loss. In this chapter we shall discuss issues relating to control of the amount of traffic entering the ATM network, in order to minimize data loss and maximize efficiency. Effective traffic management is key to the success of multi-service enterprise and public networks.

For an ATM network to provide an acceptable Quality of Service (QoS) to all the users under times of heavy loading it is essential to implement an efficient and standardized mechanism for control. If necessary, the ATM network must have the capability of scaling back the rate at which cells enter the ATM network. It should also be able to discard the cells that have already entered the network in order to reduce congestion.

Traffic management for maintaining QoS includes functions for prevention and control of congestion across an ATM network. Within ATM, traffic management can be divided into the following domains:

- Traffic Control which refers to the set of actions taken by the network to avoid congestion from ever occurring. An additional role of traffic control is to optimize the use of network resources.

- Congestion Control action taken to minimize the intensity, spread and duration of a congestion condition.

The objectives of traffic management are:

- the ability to provide user-friendly, end-to-end quality of service (QoS) guarantees for existing and new applications.

- optimal use of network resources, including efficient, dynamic bandwidth management;

- fair allocation among users and services giving higher priority to mission-critical applications;

- proactive management under congestion while maintaining network robustness and reliability.

Traffic management in ATM networks has two basic components: call admission control and congestion control. The call admission control policy controls new traffic coming into the network while the congestion control policy deals with the existing traffic sources which are already admitted into the network. The call admission control policy allows new traffic sources into the network if there is sufficient bandwidth available for the requested service; if not the new call is rejected. On the other hand, the congestion control policy controls the traffic inflow into the network by making sure that the traffic sources do not generate traffic beyond their bandwidth allowances negotiated at the time of their admission into the network.

Traffic control functions are concerned with congestion avoidance. In certain instances traffic control may fail and congestion may occur. Congestion control mechanisms are invoked at that point. These functions respond to and recover from congestion. Both fairness and stability must be guaranteed by the congestion avoidance system.

ATM forum has defined a collection of congestion control functions that operate across a spectrum of timing intervals. Table 8-1 lists these functions with respect to the response times within which they operate. The design of an optimum set of ATM Layer Traffic Controls and Congestion Controls should minimize network and end-system complexity while maximizing network utilization.

A typical ATM call goes through three basic phases with respect to ATM traffic management: call admission control, traffic contract and traffic policing phases. The call admission phase deals with the bandwidth allocation task for the user. The traffic contract phase defines the agreement established between the end user and ATM network for the traffic characteristics of the call. Finally, the traffic policing phase deals with the enforcement of the traffic contract for the duration of the call. In the following sections we will elaborate on each phase in more detail.

The problem of effectively controlling the congestion in ATM networks is currently the subject of intense research. A full-blown traffic and congestion control strategy that is widely embraced is yet to evolve. The ATM forum has published a set of traffic and congestion control capabilities aiming at simple mechanisms and realistic network efficiency (for details see ATM Forum User-Network Interface (UNI) specification 3.0). The focus in this case is on control schemes for delay-sensitive traffic such as voice and video. The subject of handling bursty traffic is a topic of ongoing research and standardization efforts.

The user is assured that the offered cell rates will meet the rate specified in the traffic contract based on the means provided by the traffic control. Traffic control ensures that the traffic contract rates are enforced such that QoS performance is achieved across all users.

This chapter describes several types of Usage Parameter Control (UPC), Call Admission Control (CAC), Priority Control (PC) and traffic shaping. Figure 8-1(a) shows the various control functions and their placements in the ATM network interfaces.

At the terminal equipment traffic shaping can be performed to ensure that the cell flows confrom with the traffic parameters. The Usage Parameter Control (UPC) function is responsible for checking (policing) the conformance of these traffic parameters at the User Network Interface (UNI). Similarly the arriving cell flows from a prior network are checked by the Network Parameter Control (NPC) functions.

As illustrated in Figure 8-1(a), functions such as Connection Admission Control (CAC), Priority Control, resource management, etc. can be employed within a particular network. Figure 8-1(b) illustrates the location of various functions with respect to the ATM switch.

Table 8-1 shows the various time scales over which a particular control function is applicable.

ITU-T Recommendation I.371 describes a set of traffic control procedures in order to maintain the required ATM network performance:

- Connection Admission Control (CAC)
- Usage Parameter Control (UPC) and Network Parameter Control (NPC)
- priority control
- traffic shaping
- network resource management
- fast resource management
- congestion control

The use of these can be adapted to a particular ATM network environment. The recommendation leaves the usage of these procedures to ATM network providers.

Table 8-1 Time scales over which a particular control function is applicable.

TRAFFIC CONTROL FUNCTIONS	CONGESTION CONTROL FUNCTIONS	RESPONSE TIME
Network Resource Management		Long Term
Connection Admission Control		Connection Duration
Fast Resource Management	Explicit Notification	Round -Trip Propagation Time
Usage Parameter Control Priority Control	Selective Cell Discarding	Cell Insertion Time

Traffic management functions are critically important to network providers and users alike, because they are the sources of enhanced QoS, operations simplicity and flexibility, and reduced network cost. In addition, they provide the potential for service differentiation and increased revenues.

I. THE TRAFFIC CONTRACT

At the time of establishing a new ATM connection, the network and the subscriber enter into a traffic contract. The network is responsible for supporting traffic at a certain level on that perticular connection, while the subscriber is responsible for not exceeding performance limits. Thus the functions related to traffic control are concerned with establishing the necessary traffic parameters and enforcing them. Once the connection is accepted, the network continues to provide the agreed upon QoS as long as the user complies with the traffic contract.

When a user requests a new Virtual Path Connection (VPC) or a Virtual Channel Connection (VCC), the user must specify the traffic characteristics in both directions for that connection. This is done by selecting a QoS from among the various QoS classes that the ATM network provides.

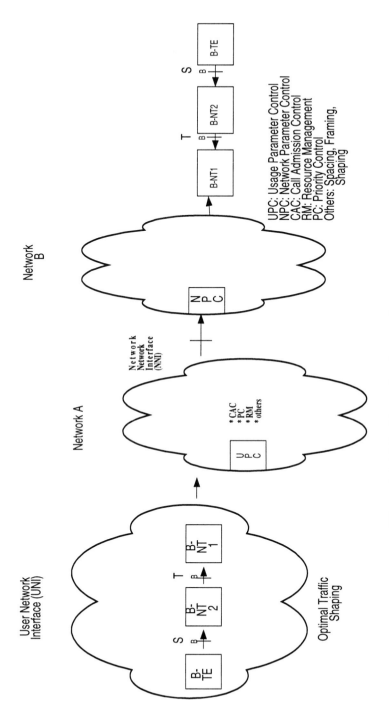

Figure 8-1 (a) Positions of various traffic control functions for an ATM switch.

UPC: Usage Parameter Control
NPC: Network Parameter Control
CAC: Call Admission Control
RM: Resource Management
PC: Priority Control
Others: Spacing, Framing,
       Shaping

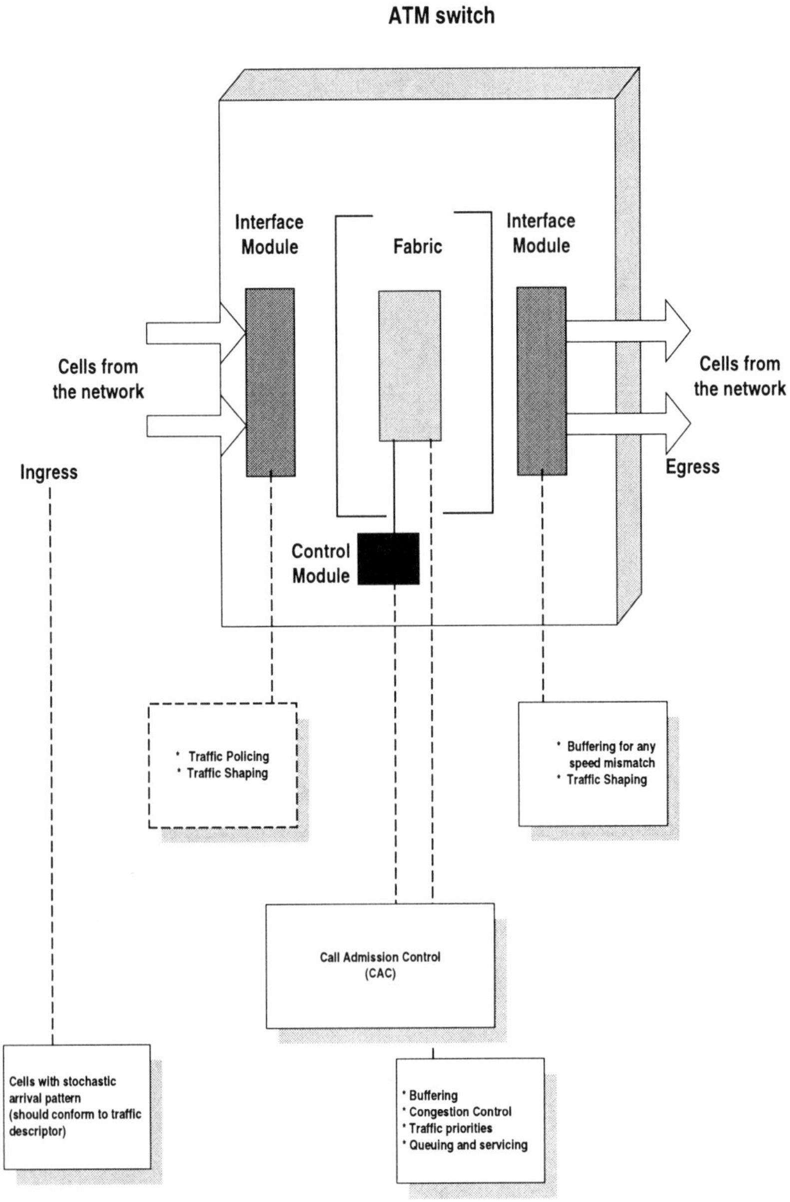

ATM switch

Figure 8-1 (b) Location of traffic management functions with respect to an ATM switch.

Before accepting the connection the network must make sure that it can commit the necessary resources for supporting the requested traffic level, without disrupting the QoS for existing connections.

A separate traffic contract exists for every VPC or VCC. This contract deals with following four interrelated aspects of any VPC or VCC ATM cell flow:

1) The traffic parameters necessary for specifying the characteristics of the ATM cell flow.

2) The expected QoS which the ATM network is expected to provide.

3) The conformance checking rules used to interpret the traffic parameters.

4) The definition of a compliant connection which the ATM network imposes on the user.

Traffic parameters describe traffic characteristics of an ATM connection. For a given ATM connection, Traffic Parameters are grouped into a Source Traffic Descriptor, which in turn is a component of a Connection Traffic Descriptor. These terms are defined as follows (ATM Forum 4.0).

Traffic Parameters A traffic parameter describes the inherent characteristics of a traffic source. It may be quantitative or qualitative. Traffic parameters may for example describe Peak Cell Rate, Sustainable Cell Rate, Burst Tolerance, and/or source type (e.g., telephone, video phone).

ATM Traffic Descriptor The ATM Traffic Descriptor is the generic list of traffic parameters that can be used to capture the traffic characteristics of an ATM connection.

Source Traffic Descriptor A Source Traffic Descriptor is a subset of traffic parameters belonging to the ATM Traffic Descriptor. It is used during the connection set-up to capture the intrinsic traffic characteristics of the connection requested by a particular source. The set of Traffic Parameters in a Source Traffic Descriptor can vary from connection to connection.

Connection Traffic Descriptor The Connection Traffic Descriptor specifies the traffic characteristics of the ATM Connection at the Public or Private UNI. The Connection Traffic Descriptor is the set of traffic parameters in the Source Traffic Descriptor, the Cell Delay Variation (CDV) Tolerance and the Conformance Definition that is used to unambiguously specify the conforming cells of the ATM connection.

Connection Admission Control procedures will use the Connection Traffic Descriptor to allocate resources and to derive parameter values for the operation of the UPC. The Connection Traffic Descriptor contains the necessary information for conformance testing of cells of the ATM connection at the UNI.

Figure 8-2 illustrates the various attributes involved in establishing a traffic contract.

Table 8-2 lists various procedures which can be used to set the values of the appropriate traffic contract parameters.

Table 8-2 Procedures used to set values of traffic contract parameters. The last two rows specify the SVCs and the PVCs, respectively.

EXPLICITLY SPECIFIED PARAMETERS		IMPLICITLY SPECIFIED PARAMETERS
Parameter values set up at circuit-setup time	Parameter values specified at subscription time	Parameter values set using default rules
Requested by user/NMS	Assigned by network operator	Assigned by network operator
Signaling	By subscription	Network operator default rules
NMS	By subscription	Network operator default rules

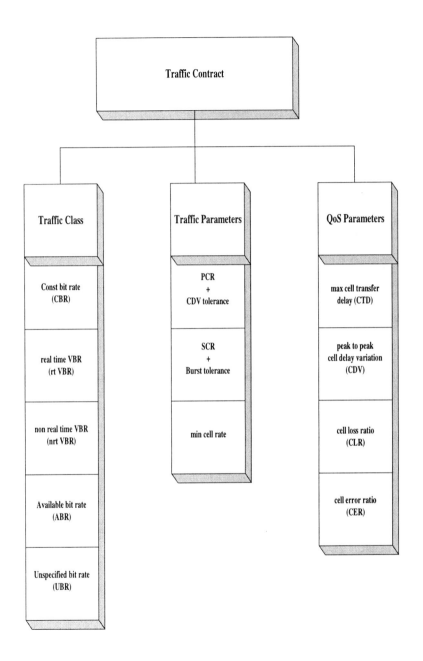

Figure 8-2   Traffic contract and its attributes.

II. BASIC QUALITY OF SERViCE (QoS) PARAMETERS

The QoS is basically defined from a perspective that is meaningful to the end user, i.e., the end workstation, customer premises network, private ATM UNI or public ATM UNI.

We can characterize end-user applications in terms of burstiness and bit rate, as well as tolerance to loss (importance), and tolerance-to-delay variation (urgency). For example:

voice traffic is very sensitive to delay variation, as is transaction traffic, and certain types of client server interactions;

video traffic is both delay and loss sensitive (though this is dependent on the type of video traffic);

image file transfer traffic is sensitive to loss;

file transfer traffic is relatively insensitive to loss and delay.

In addition, we can characterize these applications by burstiness, and average-to- peak ratios.

The QoS parameters that are of primary concern are cell loss ratio, cell transfer capacity, cell transfer delay, cell error ratio, cell misinsertion ratio and cell delay variation. They are defined as follows:

Cell loss ratio = lost cells/transmitted cells

Lost cells are those that do not reach the destination user. The **CLR** is a function of the physical link's error rate and the congestion management algorithms used in an ATM switch.

Cell error ratio = errored cells/successfully transferred cells + errored cells

Errored cells are those that arrive at the destination but contain errors in the payload.

Cell misinsertion ratio = misinserted cells/time interval

Misinserted cells are those which are not sent by the transmitting entity. These cells may arrive at the destination user either due to an undetected cell header or a configuration error.

Cell Transfer Capacity is the maximum number of successfully delivered cells occurring over a specified ATM connection during a unit of time.

Cell Transfer Delay CTD is defined as the maximum end-to-end cell transit time. The CTD is a function of the transmission delays and the ATM switch queuing delays.

It is the total time from the first bit sent from the transmitting entity until the last bit arrives at the destination. Finally, the CLR is defined as the number of lost cells divided by the number of total transmitted cells.

Figure 8-3 illustrates the probability density function of the cell transfer delay for real-time service categories. Its relation to peak-to-peak cell delay variation and maximum cell transfer delay (maxCTD) is important, since these two parameters are negotiated with the user.

Cell Delay Variation refers to the fact that some cells will be switched very rapidly through the ATM network while others may take longer due to nodal congestion, etc. In other words, CDV is the amount of time between the maximum end-to-end cell transit time and the minimum end-to-end cell transit time. CDV is a function of multiplexing many connections onto a single physical link and the variability in ATM switching queuing delays.

Since the processing delay will vary over time, the end-to-end cell transfer delay will vary over time as well. It is currently defined as a measure of how much more closely the cells are spaced than the normal interval. It can be computed at a single point against the nominal intercell spacing, or from an entry to an exit point.

It is inevitable that some variability would occur in the rate of delivery of cells due to both effects within the ATM network and at the source. In case of ATM networks, cell delay variation due to events within the network are likely to be minimal since the ATM protocol is designed to minimize processing and transmission overhead internal to the ATM network so that very fast cell switching and routing is possible. Thus the major factor for causing the variation in cell transfer delay within the ATM network is congestion. Once congestion begins, cells must be discarded; otherwise there will be a buildup of queuing delays at the affected switches.

Figure 8-4 illustrates the potential sources of cell delay variation at the source user network interface. An application may generate data at the constant bit rate for transmission across the ATM network, but processing at the three layers of the ATM model may introduce delays. At the AAL delays may be introduced during the interleaving process. At the ATM layer delay may occur while interleaving the OAM cells with the user cells.

Variations in delays during cell transmission cause problems in constant-bit-rate (CBR) services such as voice and video. For example, human perception is highly attuned to the correct corelation of audio and video. This is apparent in foreign language films that are dubbed in English. Differential delays could cause similar effects in case of the ATM network if differential delays occur causing loss of correlation between image and voice. Another example is that of applications related to file transfer. Variations in delay can result in retransmission and consequent reduction in usable throughput.

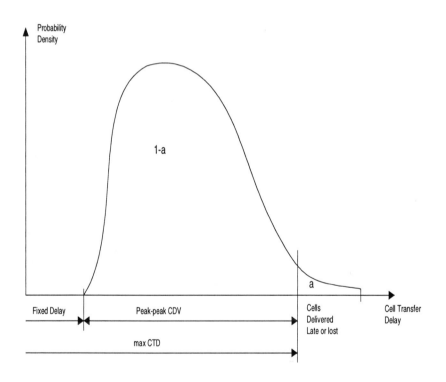

Figure 8-3 Probability density function for cell transfer delay.

In the case of voice related applications if cells arrive at larger and larger intervals or shorter and shorter intervals, the voice will sound distorted. Therefore, a constant delay accross the ATM network is absolutely crucial for providing acceptable voice services. The mechanism for providing such consistent delay on a variable delay ATM network is known as *ATM network conditioning*.

For maintaining QoS it is crucial to determine the source traffic characteristics: what exactly is required of the ATM network? The source traffic characterized by four parameters. The current ATM Forum specifications (see ATM Forum User-Network Interface (UNI) specification 3.0) define four traffic control parameters as follows:

Peak Cell Rate (PCR) places an upper bound on the amount of traffic that can be submitted for transmission over a given ATM connection.

Cell Delay Variation (CDV) At a given measurement point, it is the maximum variability observed in the pattern of cell arrivals with respect to the peak cell rate.

Consider the following example, where ATM cells are represented by small automobiles.

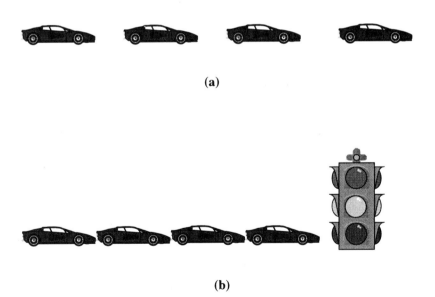

(a)

(b)

Figure 8-4 (a) Traffic with constant spacing of vehicles; (b) effect of a traffic signal on the traffic pattern.

Constant bit-rate traffic is shown in Figure 8-4(a), where the time/spacing between cells is constant.

This perfect traffic can be spoiled by the first ATM switch where decisions have to be made about which cell goes first, illustrated here in Figure 8- 4(b) by a traffic light.

The traffic light, or ATM switch, can cause the cells to bunch. Left uncorrected, the cells will arrive at the destination in bunches and will thus show variations in their individual journey times. This is called CDV, cell delay variation.

This problem can be corrected, by doing active traffic shaping (discussed in detail in Section 8-6), after the last switch, to restore the original cell pacing, as illustrated in Figure 8-5. This is sometimes achieved for guaranteeing the required QoS at the cost of increasing the total delay (CTD – cell transmission delay).

Figure 8-5 Traffic pattern after traffic shaping.

Sustainable Cell Rate (SCR) For a given duration of the connection, it defines the upperbound on the average rate at which cells are transferred through an ATM connection.

Maximum Burst Size (MBS) At a given measurement point, it is the maximum variability observed in the pattern of cell arrivals with respect to the sustainable cell rate.

The first two are relevant for the constant-bit-rate (CBR) sources, while all four may be used for variable-bit-rate (VBR) sources. QoS parameters are defined at measurement points which coincide with interfaces shown in Figure 8-6(a). As you can see, the switch congestion management and queuing algorithms play a critical role in the performance an ATM network can deliver. Figure 8-6(b) graphically depicts the three negotiated QoS parameters: the Cell Transfer Delay (CTD), the Cell Delay Variation (CDV), and the Cell Loss Ratio (CLR).

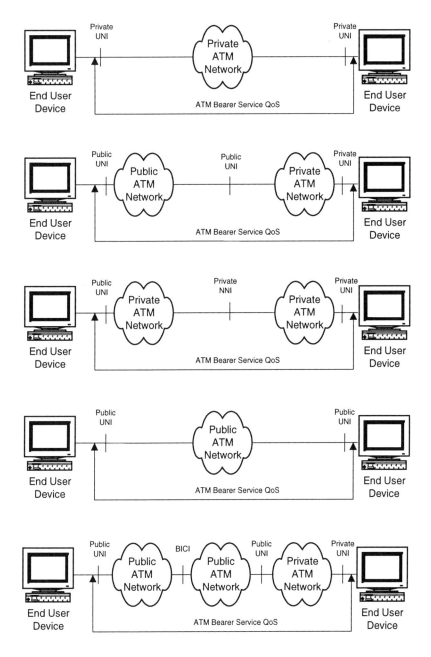

Figure 8-6 (a) QoS parameters are defined at measurement points which coincide with interfaces.

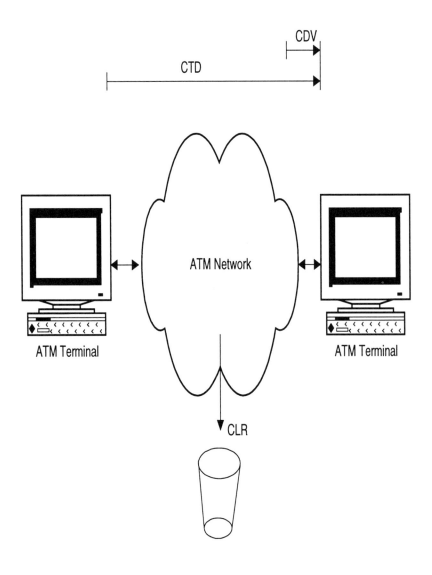

Figure 8-6 (b) The three negotiated QoS parameters and their relationship to the ATM network.

Table 8-3 describes how the three QoS parameters are related to the various types of traffic that could be transported over an ATM network. Only three of these parameters are defined as negotiable parameters by the ATM Forum; the others defined earlier in this section are not negotiated.

Table 8-3 Traffic types and QoS parameters.

TRAFFIC TYPE	CTD	CDV	CLR
Voice calls	Very sensitive: medium delays require echo cancelers and large delays are unusable	Very sensitive: larger CDV translates into larger buffers at the destination and larger CTD	Moderately sensitive: dropped data are not retransmitted but quality of the communication is reduced
Video conferen-cing	Very sensitive: large delays are unusable	Very sensitive	Moderately sensitive
Video on demand	Moderately sensitive: source should be able to respond to remote control commands	Very sensitive	Moderately sensitive
Data	Insensitive: connections have long time outs and large retransmission windows	Insensitive: destination usually has large buffers	Very sensitive: whole packets (lots of cells) need to be retransmitted even if one cell is lost

The ATM Forum has defined a number of traffic control functions to maintain QoS of ATM connections (see ATM Forum 4.0: Traffic Control for details). The following generic functions form a framework for managing and

143

controlling traffic and congestion in ATM networks and may be used in appropriate combinations.

Network Resource Management (NRM): Provisioning may be used to allocate network resources in order to separate traffic flows according to service characteristics. Virtual paths are a useful tool for resource management.

Connection Admission Control (CAC) is defined as the set of actions taken by the network during the call setup phase (or during call re-negotiations phase) in order to determine whether a virtual channel/virtual path connection request can be accepted or should be rejected (or whether a request for re-allocation can be accommodated). Routing is part of CAC actions.

Usage Parameter Control (UPC) is defined as the set of actions taken by the network to monitor and control traffic, in terms of traffic offered and validity of the ATM connection, at the end-system access point. Its main purpose is to protect network resources from malicious as well as unintentional misbehavior, which can affect the QoS of other already established connections, by detecting violations of negotiated parameters and taking appropriate actions.

Priority Control For some service categories the end-system may generate different priority traffic flows by using the Cell Loss Priority bit. A network element may selectively discard cells with low priority if necessary to protect, as far as posible, the network performance for cells with high priority.

Traffic Shaping Traffic shaping mechanisms may be used to achieve a desired modification of the traffic characteristics.

Feedback Controls are defined as the set of actions taken by the network and by the end-systems to regulate the traffic submitted on ATM connections according to the state of network elements.

Other generic control functions are for further study.

III. CONNECTION ADMISSION CONTROL

Call admission control defines a set of rules or algorithms to decide whether a new call should be accepted or not. The decision is made based on the traffic characteristics of the new call and availability of required network resources to handle the additional traffic, without affecting the resource requirements of existing calls. In other words, the additional potential traffic to be generated by the new call should not have any effect on the traffic performance requirements of existing calls.

ATM connection admission control (CAC) is performed when a connection request has been made. Such a connection may be set up either at service provision time on a semipermenant basis or by means of a signalling protocol at the call origination time on a dynamic basis. The CAC mechanism must determine if it can set aside the proper amount of resources, such as resources necessary to provide adequate bandwith or a bounded transmission delay, for servicing that connection. If CAC mechanism determines that the necessary resources cannot be allowed, the ATM network will not accept that connection request.

When a connection request is made the local network node at the source end of the ATM network may have the necessary resources. It is also possible that the local network node at the destination may also have the necessary resources. However, the connection cannot be accepted at all if for all the possible paths through the internal or backbone ATM, the ATM network nodes cannot provide the requested QoS.

There are several CAC algorithms described in the literature. We will briefly describe some of the major CAC algorithms.

CAC is actually a software function in a switch that is invoked at the call setup time, when a VCC or a VPC is established. It accepts the request only if the QoS for all existing conections will still be met if the request is accepted. CAC can be performed either on a node-by-node basis or in a centralized system. If the request is accepted the CAC must determine the UPC/NPC parameters, routing and allocate resources such as trunk bandwith, buffer space and internal switch resources.

For supporting the high switching rates required for ATM, CAC must be simple and rapid. Figure 8-7 shows a simple CAC algorithm which is based on peak rate allocation. The algorithm operates by determining   whether the sum of the peak rates will exceed the trunk bandwidth. If so, the connection request is denied.

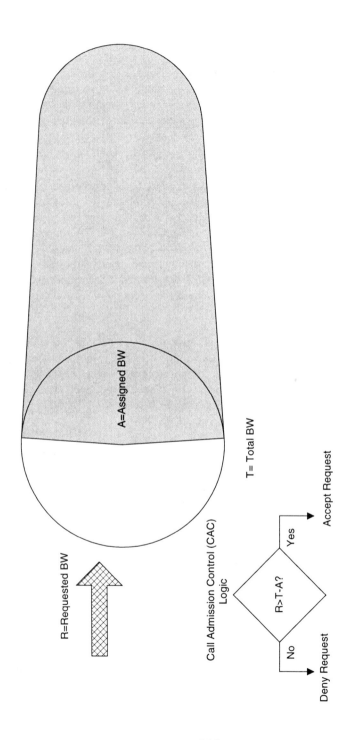

Figure 8-7 Call Admission Control algorithm.

Figure 8-7 shows a pipe for transmission which has a bandwidth of X. It can be seen that a certain portion of it, namely Y, is already assigned to the existing connections. The user request for bandwidth B is illustrated on the left in Figure 8-7. The request is submitted to CAC logic based on the peak rate. The logic of the algorithm operates by comparing the requested bandwidth B to the available bandwidth (X-Y). If B exceeds the available bandwidth, the request is denied; otherwise it is accepted.

The complexity of the CAC algorithm is related to the traffic descriptor. A connection request defines the source traffic parameters and the requested QoS class. Table 8-4 depicts another sample CAC algorithm based on the type of service category.

Table 8-4 Sample call admissions control algorithm.

SERVICE CATEGORY	BANDWIDTH ALLOCATION ALGORITHM	EXCESS BANDWIDTH AVAILABLE FOR ABR/UBR
High QoS CBR	Allocate two times PCR	PCR
Medium QoS CBR	Allocate PCR	None
High QoS rtVBR	Allocate PCR	PCR-SCR
Medium QoS rtVBR	Allocate 1.5*SCR	0.5*SCR
NrtVBR	Allocate SCR	None
ABR	Allocate MCR + 0.01% of link rate	None
UBR	Allocate 0.01% of link rate	None

Source traffic can be characterized by several parameters which address the issue of what exactly is required of the ATM network? Four important parameters of the traffic characterization are:

1. **Average bit rate:** Arithmetic mean over time for bit rate at which the source operates.

2. **Peak bit rate:** Highest bit rate the source is capable of sending.

3. **Physical bit rate:** The actual bit rate provided by the link from the user to the local ATM network node.

4. **Peak duration:** A measure of how long the source is capable of maintaining the peak bit rate.

The ratio of peak to average bit rate is referred to as burstiness (burst ratio), which is vital information at the time of establishing a connection. Burstiness can vary from one in the case of voice to one hundred (for compressed video) or even a thousand for Local Area Network related data.

Based on these QoS parameters declared by the user the CAC can determine the type of service required. For example if all three bit rates are declared as zero, a connection of the type available-bit-rate (ABR) service is provided by the network. In this case the connection is assigned whatever the bit rate is available at that particular time.

It is also very important to know the peak duration before the connection is established. It allows the ATM network node to determine the maximum number of cells that are likely to enter the network from a given connection during a particular time interval. It is important to note that the ATM Forum proposals allow for all the parameter values detailed above to be negotiated. Thus, if a connection request for a peak bit rate of ten Mbps is rejected the user can try again with a revised bit rate of five Mbps.

A practical example of the functioning of the CAC mechanism can be provided by taking a look at a typical ATM switch such as MainStreetXpress 36190 from Siemens Telecom Networks. MainStreetXpress 36190 converts Connection of Service requests signaled across the UNI into the following VC parameters:

service category: constant, variable, unspecified, available bit rate (CBR, VBR, UBR, ABR)

traffic descriptors: peak, sustained, minimum cell rates (PCR, SCR, MCR), cell delay variation tolerance (CDVT), maximum burst size (MBS)

QoS objectives: cell-loss ratio (CLR), cell transit delay (CTD), cell delay variation (CDV)

All of these parameters are the inputs to CAC algorithms.

CAC operates using different algorithms under normal and heavy traffic conditions. In addition, different algorithms are used for constant and variable rate VCs. The end result is acceptance or rejection of the connection request, or potentially, negotiation with the user once the ability to negotiate is standardized.

Typically CAC may permit a certain amount of resource overbooking in order to increase statistical multiplex gain. Other factors, such as the slack implied by the compliant connection definition, CDV tolerance parameter, buffer size available for a certain cell loss, delay in QoS objective, etc. may also allow for a relaxation in the admission threshold of the CAC algorithm.

IV. USAGE/NETWORK PARAMETER CONTROL (UPC/ NPC)

UPC/NPC is commonly referred to as policing since they play a role similar to that of a police in the society. UPC/NPC are responsible for fair treatment of all users according to their traffic contracts. Once a connection has been accepted by the connection admission control function, these two parameter control functions ensure fair treatment by making sure that the bandwidth and the buffering resources are fairly allocated among all users.

Usage Parameter Control is defined as the set of actions taken by the network to monitor control traffic in terms of traffic offered and validity of the ATM connection, at the user access. Its main purpose is to protect network resources from malicious as well as unintentional misbehavior which can affect the QoS of other already established connections by detecting violations of negotiated parameters and taking appropriate actions.

These control functions make sure that a single user does not "hog" all the available resources. In other words, these parameter control functions protect the network resources from an overload on one connection that would adversely affect the other connections. This is accomplished by using various algorithms that detect violations of assigned parameters and take appropriate actions.

The control functions can be performed at both the virtual path and the virtual channel levels; however, the exact location depends on the configuration. Connection monitoring encompasses all connections crossing the Public UNI. Usage Parameter Control applies to both user VCCs and VPCs and signalling virtual channels. Methods for monitoring meta-signalling channels and OAM flows are defined by the ATM forum for further study.

The monitoring task for usage parameter control is performed for VCCs and VPCs respectively by the following two actions:

1. checking the validity of VPI and VCI (i.e., whether or not VPI/VCI values are associated with active VCCs) and monitoring the traffic entering the network from active VCCs in order to ensure that parameters agreed upon are not violated;

2. checking the validity of VPI (i.e., whether or not VPI values are associated with active VPCs), and monitoring the traffic entering the network from active VPCs in order to ensure that parameters agreed upon are not violated.

UPC is performed at the points where the VP or VC link connected to the Network Terminator (NT), i.e., user's PC, etc., is terminated in the network. Figure 8-8 illustrates the three possibile configurations.

In Figure 8-8 the following notation is used:

VC-Sw stands for Virtual Channel Switching Function;

VP-Sw stands for Virtual Path Switching Function;

VC-Sws and VP-Sws refer to the first switches on the public network side of the Public UNI or on the network side of the private UNI;

VC-Sws and VP-Sws refer to the first switches on the private network side of the public UNI when UPC is implemented in private network elements;

A VC-Sw or a VP-Sw may, respectively, be a VC or VP concentrator.

As shown in CASE A of Figure 8-8, the UPC function shall be performed within the VC-Sw on VCCs before the switching function is executed, if the NT is connected directly to a VC-Sw.

As shown in CASE B of Figure 8-8, the UPC function shall be performed within the VP-Sw on VPCs only and within the VC-Sw on VCCs only, if the NT is connected directly to VC-Sw via VP-Sw.

For CASE C of Figure 8-8, the UPC function shall be performed within the VP-Sw on VPCs only if the NT is connected to an end-system or to another network via VP-Sw. In CASE C of Figure 8-8, the VCC usage parameter control will be done by the first public network (if any) where VC-Sw is present.

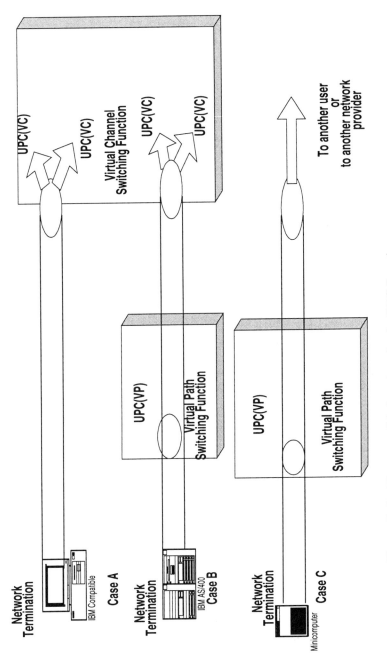

Figure 8-8 Three possible locations for the UPC Functions.

151

The traffic parameters that may be subject to UPC enforcement are those included in the source traffic descriptor. Thus, the PCR of the CLP = 0 + 1 cell flow is required to be subject to UPC for all types of connections at the public UNI. Even when the parameters are specified, the enforcement of SCR and MBS is network specific.

The parameter control functions use various algorithms like the Peak Cell Rate Algorithm, Substitute Cell Rate algorithm, etc. to determine whether the user is complying with the traffic contract. For example, a traffic flow is determined to be compliant if the peak rate of the cell transmission does not exceed the agreed peak cell rate, subject to the possibility of cell delay variation within the agreed bound.

A number of desirable features of the User Parameter Control algorithm can be identified as follows (ATM Forum: Traffic Management Specification 4.0):

1. capability of detecting any non-compliant traffic situation;

2. selectivity over the range of checked parameters (i.e., the algorithm could determine whether the user's behavior is within an acceptance region);

3. rapid response time to parameter violations;

4. simplicity of implementation.

There are two sets of requirements relating to the UPC:

1. those which relate to the quality of service impairments the UPC might directly cause to the user cell flow;

2. those which relate to the resource the operator should allocate to a given connection and the way the network intends to protect those resources against misbehavior from the user side (due to fault conditions or maliciousness).

Methods for evaluating UPC performance and the need to standardize these methods are for further study. The following two performance parameters are identified that could be considered when assessing the performance of UPC mechanisms.

Response time: the time to detect a given non-compliant situation on a VPC/VCC under given reference conditions.

Transparency: for given reference conditions, the accuracy with which the UPC initiates appropriate control actions on a non-compliant connection and avoids inappropriate control actions on a compliant connection.

Generic Cell Rate Algorithm (GCRA) is used to determine conformance with respect to the traffic contract (for a detailed discussion on GCRA refer to the ATM forum: Traffic Management Specification 4.0). For each cell arrival, this algorithm determines whether the cell conforms to the traffic contract of the ATM connection.

Although traffic conformance is defined in terms of the GCRA, the network is not required to use this algorithm (or the same parameter values) for the UPC. Rather, the network may use any UPC approach as long as the operation of the UPC supports the QoS objectives of a compliant connection. Thus the UPC function is not required to implement the exact form of the GCRA; however it, must implement one or more equivalent algorithms to enforce conformance.

The GCRA is a virtual scheduling algorithm or a continuous-state Leaky Bucket Algorithm as defined by the flowchart in Figure 8-9. The GCRA is used to define, in an operational manner, the relationship between the following set of parameters:

1) the peak cell rate (PCR) and the cell delay variation tolerance,

2) the sustainable cell rate (SCR) and the Burst Tolerance (BT).

In addition, the GCRA is used to specify the conformance of the declared values of the above two tolerances, as well as declared values of the traffic parameters PCR and SCR and maximum burst size (MBS). The GCRA is used to specify the conformance, at either the public or private UNI (for details see ATM Forum: Traffic Management Specification, Section 4.4.3.2 and Annex C.4, respectively).

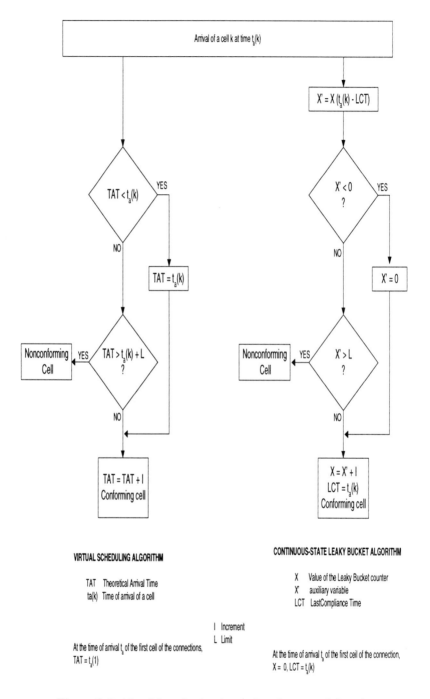

Figure 8-9 Algorithms for implementing the control functions.

As illustrated in Figure 8-9 the GCRA is defined with two parameters:

1) **Increment (I):** the increment parameter corresponds to the inverse of the compliant rate, i.e., the fill rate of the bucket.

2) **Limit (L):** the limit parameter corresponds to the number of cells that can burst at a higher rate, i.e., the size of the bucket.

Note: I and L are not restricted to integer values.

Thus, the notation "GCRA(I, L)" means the Generic Cell Rate Algorithm with the value of the increment parameter set equal to I and the value of the limit parameter set equal to L.

The GCRA is formally defined in Figure 8-9, which is a generic version of the GCRA algorithm defined in the ATM Forum documentation (i.e., Figure 1 in Annex 1 of I.371 draft). It illustrates two distinct algorithms referred to as the Virtual Scheduling Algorithm and the Continuous Leaky Bucket Algorithm.

The virtual scheduling algorithm updates a Theoretical Arrival Time (TAT), which is the "nominal" arrival time of the cell assuming that the active source sends equally spaced cells. If the actual arrival time of a cell is not "too" early relative to the TAT; in particular, if the actual arrival time is after TAT - L, then the cell is conforming; otherwise the cell is nonconforming.

The continuous-state leaky bucket algorithm can be viewed as a finite-capacity bucket whose real-valued content drains out at a continuous rate of I unit of content per time-unit and whose content is increased by the increment I for each conforming cell. Equivalently, it can be viewed as the work load in a finite-capacity queue or as a real-valued counter. If, at a cell arrival, the content of the bucket is less than or equal to the limit value, L, then the cell is conforming; otherwise the cell is nonconforming. The capacity of the bucket (the upper bound on the counter) is L + 1.

When more than one traffic descriptor is used (for example, PCR and SCR) for a connection, multiple leaky buckets are cascaded, with the highest rate being policed first. For example, ATM connections carrying frame relay service may actually define three traffic descriptors and thus require three leaky buckets: PCR0, PCR1 and SCR0. Figure 8-10 illustrates an example of traffic policing with two leaky buckets.

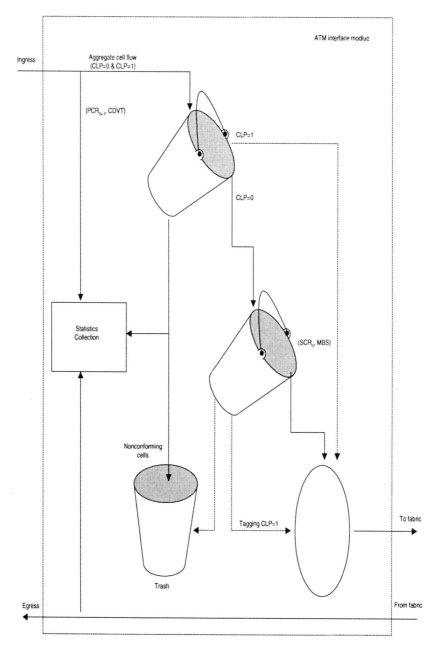

Figure 8-10 Implementation of traffic control using two leaky buckets.

The two algorithms in Figure 8-9 are equivalent in the sense that for any sequence of cell arrival times, the two algorithms determine the same cells to be conforming and thus the same cells to be nonconforming.

The two algorithms are easily compared by noticing that at each arrival epoch, ta(k), and after the algorithms have been executed, as in Figure 8-9,

$$TAT = X + LCT.$$

Tracing the steps of the virtual scheduling algorithm in Figure 8-9, at the arrival time of the first cell ta(1), the theoretical arrival time TAT is initialized to the current time, ta(1). For subsequent cells, if the arrival time of the kth cell, ta(k), is actually after the current value of the TAT, then the cell is conforming and TAT is updated to the current time ta(k), plus the increment 1.

If the arrival time of the kth cell is greater than or equal to TAT - L but less than TAT (i.e., as expressed in Figure 8-9, if TAT is less than or equal to ta(k) + L), then again the cell is conforming, and the TAT is increased by the increment 1.

Lastly, if the arrival time of the kth cell is less than TAT-L (i.e., if TAT is greater than ta(k) + L), then the cell is nonconforming and the TAT is unchanged.

Similarly, tracing the steps of the continuous-state leaky bucket algorithm in Figure 8-9 at the arrival time of the first cell ta(1), the content of bucket, X, is set to zero and the last conformance time (LCT) is set to ta(1).

At the arrival time of the kth cell, ta(k), first the content of the bucket is provisionally updated to the value X, which equals the content of the bucket, X, after the arrival of the last conforming cell minus the amount the bucket has drained since that arrival, where the content of the bucket is constrained to be non-negative.

Second, if X is less than or equal to the limit value L, then the cell is conforming, and the bucket content X is set to X' plus the increment I for the current cell, and the last conformance time, LCT, is set to the current time ta(k). If, on the other hand, X' is greater than the limit value L, then the cell is nonconforming and the values of X and LCT are not changed.

UPC Actions (Cell Tagging and Discard)

The UPC is intended to ensure conformance by a connection with the negotiated traffic contract. The objective is that a connection will never be able to exceed the traffic contract.

At the cell level, actions of the UPC function may include:

1) cell passing;

2) cell tagging (network option); cell tagging operates on CLP=0 cells only, by overwriting the CLP bit to 1;

3) cell discarding.

An optional cell tagging function exists for both the network and the user. The availability of tagging function depends on the conformance definition of the connection. If supported by the ATM network, the user may select this tagging function when the connection is established.

If the tagging option is used for a connection, cells with CLP=0, which are identified by the UPC function to be nonconforming to the CLP=0 cell stream, are converted to CLP=1 cells. Those cells that have the CLP bit converted from '0' to '1' are called tagged cells.

The algorithm functions as follows:

If a tagged cell is identified by the UPC function to be conforming to the CLP = 0 + 1 cell stream, then it is passed; otherwise, it is discarded.

If a cell is submitted by an end-system with CLP = 1 and it is identified by the UPC function to be conforming to the CLP = 0 + 1 cell stream, then it is passed; otherwise, it is discarded.

Thus, tagging is used to transform a nonconforming CLP = 0 cell into a conforming CLP = 1 cell when the cell is conforming to the aggregate CLP = 0 + 1 flow.

If a CLP = 0 cell is not conforming to the aggregate CLP = 0 + 1 flow, tagging will not make it conforming. Therefore, the tagging option never applies to the aggregate CLP = 0 + 1 flow, but applies to the CLP = 0 flow.

As a result, cells are passed when they are identified by the UPC as conforming. Thus, if a tagged cell is passed, then it is considered to be conforming. On the other hand, cells are discarded when they are identified by the UPC as nonconforming.

In some cases, following the UPC function, traffic shaping may be used to perform cell re-scheduling (e.g., to reduce cell clumping) on cells identified by the UPC as conforming. In addition to the above actions at the cell level, the UPC function may initiate the release of an identified non-compliant SVC.

Enforcement of the PCR bound by the usage parameter control allows the ATM network to allocate sufficient resources to ensure that the network performance objectives, i.e., cell loss ratio, etc., can be achieved. Figure 8-11 illustrates the equivalent-terminal configuration for a PCR reference model. It provides an abstract framework for describing the UPC at the public UNI. The virtual shaper provides conformance with the peak transmission interval T. The cell delay variation is represented by τ and τ^* in Figure 8-11. The virtual shaper provides shaping with conformance to GCRA(T,CDV). At the Phy-Layer SAP, CDV is zero. Thus it applies GCRA(T,0). At the private UNI CDV is τ; thus GCRA(T, τ) is applicable. The CDV at the public UNI is τ^*; hence GCRA(T, τ^*) is applicable.

As shown in Figure 8-11 the traffic sources, the multiplexer (MUX) and the virtual shaper define the equivalent-terminal. Each traffic source generates requests to send ATM cells at its own rate. The virtual shaper is intended to reflect some smoothness in the cell flow offered to the ATM connection.

Figure 8-12 illustrates the SCR and BT reference model as defined by the ATM forum: Traffic Management Specification 4.0. The virtual shaper provides conformance with the peak transmission interval Ts.

In this case the burst tolerance is added to the CDV value before using it as a parameter for the GCRA. In this case the virtual shaper provides shaping with conformance to GCRA(Ts , BT + CDV).

At the Phy-Layer SAP, again the CDV is zero. Thus it applies GCRA(Ts, BT). At the private UNI CDV is τ; thus GCRA(Ts, BT + τ) is applicable. The CDV at the public UNI is τ^*; hence GCRA(Ts, BT + τ^*) is applicable.

159

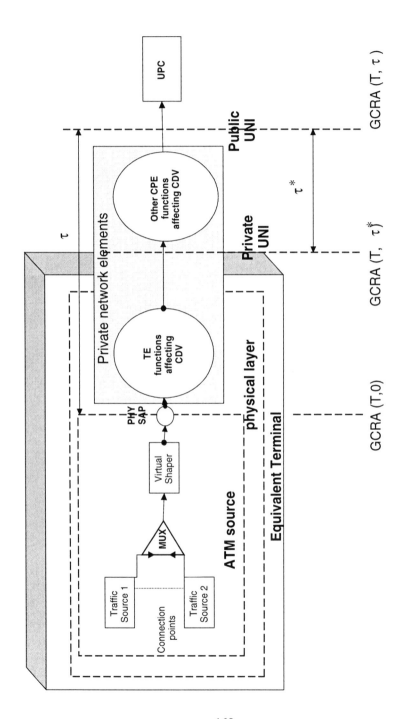

Figure 8-11 Enforcement of Peak Cell Rate.

160

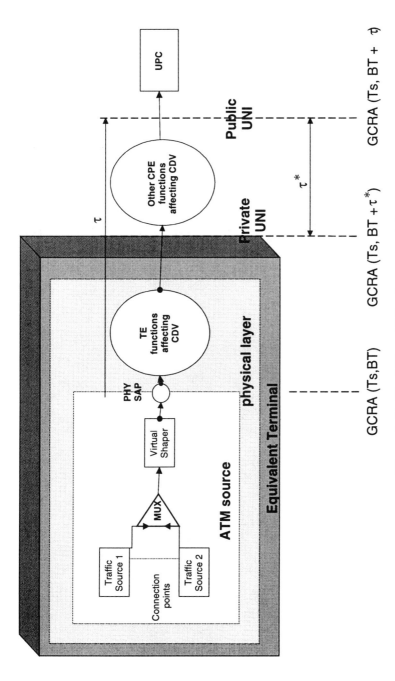

Figure 8-12 The SCR and BT reference model.

161

V. ATM SERVICE CATEGORIES

Networks must be designed to maximize the utilization of resources such that the best possible cost to performance ratio is attained while meeting user service expectations. ATM standards and implementation agreements specify many capabilities for meeting these goals, while other capabilities are left to vendors and network providers to implement using proprietary technology.

Traffic management is based on the service category and associated quality of service that a user or group of users requires on a connection. The service contract for a connection is determined according to whether the connection is permanent or switched.

Permanent virtual connections: For permanent virtual connections (virtual path or virtual channel), the service contract is defined at the user endpoints and along the route that has been selected for the connection using the provisioning capabilities of the network vendor.

For example, Siemens's 46020 Network Management System (NMS) instructs the appropriate network elements to establish the permanent virtual connection according to the traffic contract parameters provided.

Switched virtual connections: Switched virtual connections enable the network to allocate bandwidth on demand. End users make their service request to the network using functions provided by the end system for accessing the ATM user-to-network interface.

Once the UNI switched virtual connection request is made, the network must determine if the connection request can be accepted and then route the connection appropriately.

As discussed in detail in Chapter 6, four service categories are specified by the ATM Forum:

constant bit rate (CBR)
variable bit rate (VBR)
available bit rate (ABR)
unspecified bit rate (UBR)

Variable bit rate has two variants or subclasses: real-time VBR and non-real-time VBR.

Each service category has several attributes, or parameters, associated with it. Some are traffic parameters, which describe the inherent characteristics of a connection's traffic source. Others are called quality-of-service (QoS) parameters, which define performance attributes required by an application. In the ATM Forum's Version 4.0 traffic management specification, some QoS parameters are negotiable via signaling, whereas others are non-negotiable.

Constant bit-rate (CBR) service is the highest priority service category, designed for isochronous traffic that must meet strict throughput and delay requirements. CBR traffic is not bursty. The parameter used for CBR is the Peak Cell Rate (PCR). The bandwidth is the same under all circumstances. Once a traffic contract has been negotiated, the transmission of cells within the PCR is guaranteed by the ATM network. CBR is used for voice and video transmission.

As illustrated in Figure 8-13, a bit stream originating at a source must be able to be reconstructed from cells at the destination within the constraints of the CBR connection attribute values to ensure that the required quality of the received bit stream is attained.

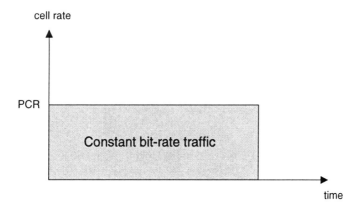

Figure 8-13 CBR traffic.

Examples of CBR traffic are telephone voice, interactive video, video (television) and audio (radio) distribution, and emulation of digital circuits such as T1 and DS3.

Applications whose information transfer is bursty can utilize one of two variable bit-rate (VBR) options: real-time VBR as defined in UNI 3.1, or non-real-time VBR as defined in the ATM Forum Version 4.0 traffic management specification.

Variable bit-rate traffic (VBR)   VBR is bursty. Parameters used are Peak Cell Rate (PCR), Sustainable Cell Rate (SCR) and Maximum Burst Size (MBS). Once a traffic contract has been negotiated, the transmission of cells within the VBR parameters is guaranteed by the ATM network. The number of cells allowed to exceed SCR is set by MBS. A cell rate which exceeds the PCR is permissible.

Figure 8-14 illustrates the relationship between the cell rate and the time for a VBR traffic source.

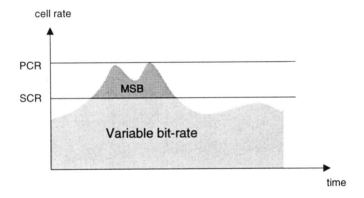

Figure 8-14 VBR traffic pattern.

For example, a voice connection that utilizes functions such as voice compression and silence suppression can be implemented using real-time VBR instead of CBR to save bandwidth while continuing to meet delay constraints. Examples of non-real-time VBR applications are airline reservations, banking transactions, process monitoring, and frame relay interworking.

Traffic that requires no service guarantees can utilize the unspecified bit-rate (UBR) service category.

Unspecified Bit Rate (UBR)   There are no parameters for UBR. The transmission of cells is not guaranteed by the ATM network. UBR is sometimes referred to as a best-effort service.

Figure 8-15 illustrates this type of traffic pattern.

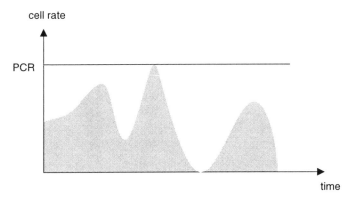

Figure 8-15 UBR type of traffic pattern.

Examples of applications that can use "background" bandwidth allocated to UBR traffic are image information such as news and weather pictures, LAN interconnection, telecommuting, file transfer and electronic mail.

As discussed in Version 4.0 of the traffic management specification, a user can specify two attributes for a UBR connection: peak cell rate and cell delay variation tolerance. Network vendors may optionally choose to subject UBR traffic to connection admission controls and usage parameter controls. UBR is a good way to utilize excess bandwidth for applications that will occasionally have cells discarded without serious consequence, such as TCP/IP-based applications which are able to recover packets using routine re-transmission protocols.

The ATM Forum has specified a service category called the Available Bit Rate (ABR) for more efficiently and fairly managing excess bandwidth capacity.

Available bit-rate traffic (ABR)   The parameters required to specify an ABR service are Minimum Cell Rate (MCR) and PCR which is necessary to limit the peak cell-rate. The area between MCR and PCR is known as the Allowed Cell Rate (ACR). The ACR varies, depending upon the current network situation.

ABR is designed for applications that do not have rigorous cell transfer delay tolerances but do have low cell-loss requirements such as distributed computing applications. ABR applications, which compete fairly for excess capacity through flow control algorithms, specify minimum throughput requirements. Figure 8-16 illustrates an ABR traffic pattern.

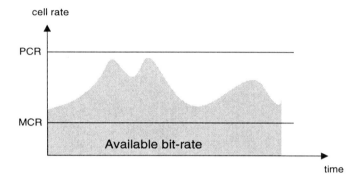

Figure 8-16 ABR traffic pattern.

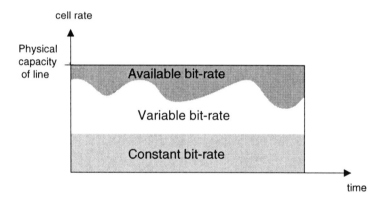

Figure 8-17 Various types of traffic patterns with respect to the physical line.

Examples of ABR applications include existing UBR applications (news and weather pictures, LAN interconnection, telecommuting, file transfer and electronic mail) that require more predictable behavior. Additional examples are defense applications and banking applications with critical data transfer requirements. Such use of three types of traffic over ATM optimizes usage of capacity of the physical link.

Figure 8-17 shows the relationship between different types of traffic patterns. Table 8-5 describes how the Service Categories are related to the various types of traffic that could be transported over an ATM network.

Table 8-5 Traffic types associated with each service categories.

TRAFFIC TYPE	CBR	UBR	rt-VBR	nrt-VBR	ABR
Data	x	x	x	x	x
Compressed voice, compressed video (MPEG)	x	-	x	-	-
Voice, video	x	-	-	-	-

Table 8-6 shows the parameters that are specified for each of the defined service categories. Both voice and video require the QoS parameters that are specified for CBR and rtVBR traffic. Data depending upon the application may or may not require the specification of QoS parameters.

VI. TRAFFIC SHAPING

When traffic does not conform to the terms of a traffic contract, the traffic can be routed through a shaping function to bring it into conformance with the contract. For example, real-time VBR traffic whose peak rate bursts are higher than the contracted peak cell rate can be modified to extend the duration of the burst by spreading the cells in time as long as the burst tolerance and other parameters associated with the contract remain valid.

Figure 8-18 illustrates the effect of traffic shaping on a stream of bursty cells.

It is very useful to complement UPC with corrective traffic shaping. This can guarantee correct transmission of the user traffic and reduce costs by reducing the peak cellrate required.

167

Table 8-6 Parameters associated with each service categories.

TRAFFIC TYPE	CBR	UBR	rt-VBR	nrt-VBR	ABR
QoS Parameters	-	-	-	-	-
CDV	x	-	x	-	-
CTD	x	-	x	-	-
CLR	x	-	x	x	x
Traffic Parameters	x	x	-	-	-
PCR	x	-	x	x	-
SCR, MBS	-	-	x	x	-
MCR	-	-	-	-	x
Feedback	-	-	-	-	x

For traffic shaping, cell transmission delay (CTD) and the cell delay variation (CDV) are important parameters. It is possible to trade off CTD against CD, for example, more CTD for less CDV during the traffic shaping process. The sources of CTD and CDV are the ATM switches and the ATM access concentrators.

Figures 8-19 and 8-20 illustrate how the parameters PCR, SCR, MBS describe the quality of a connection. Traffic violating these parameters will be marked for preferred discarded later on (CLP).

168

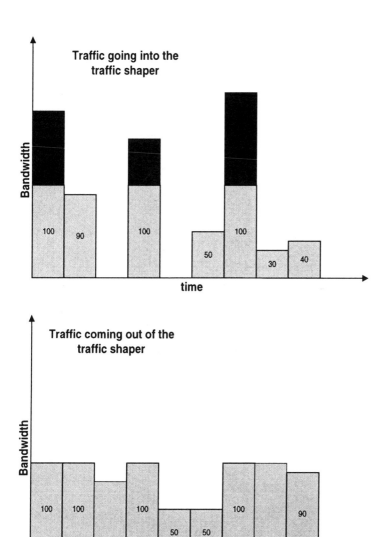

Figure 8-18 Effect of traffic shapping.

169

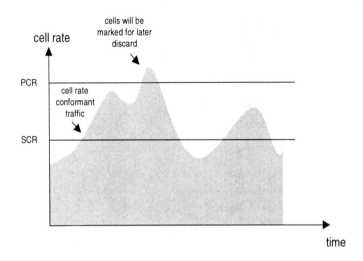

Figure 8-19 Traffic parameters and traffic shaping.

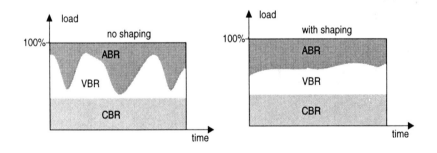

Figure 8-20 Shaping for the three services.

The traffic shaping function modifies the ATM traffic flow to achieve improved network efficiency, for interworking to ATM switches with small

170

buffers and for virtual paths (VPs) originating at the ATM switch. Traffic shaping can be done either at the ingress or egress side of an ATM switch.

Traffic shaping at the ingress side is used to modify the traffic flow of terminal equipment and ATM switches that are not able to keep the traffic contract, for instance, sending ATM cells with a higher peak bit rate than allowed. Traffic shaping adapts the traffic characteristics of the ATM cell stream to the traffic contract. This leads to a more efficient use of network resources by allowing network operators to allocate only those resources required according to the traffic contract.

Traffic shaping at the egress side of the ATM node is provided to protect subsequent ATM switches with small buffers from data bursts leading to buffer overflow. In addition, egress shaping is required to ensure that VPs originating at the ATM switch conform to the traffic contract.

A CBR source shapes traffic using PCR. Thus, it schedules a cell every (1/PCR) unit of time. A VBR source shapes bursts (MBS) at the PCR rate, while ensuring that the overall SCR is maintained.

An ABR source requires dynamic shaping. In order to comply to the flow control mechanism used to support the ABR service, the ABR source traffic shaper varies according to the feedback messages it receives from the network and the application's requirements. Note, the source does not have to be the connection endpoint. To shorten feedback loops in networks with large delays the use of virtual sources and virtual destinations can be set up within the ATM switches. These behave exactly like ABR sources and destinations. Figure 8-21 illustrates the various applications and the service types.

VII. FLOW CONTROL AND CONGESTION CONTROL

Congestion control allows the management of traffic when many users contend for finite network resources. Congestion management is one of the most important functions that an ATM switch will have to perform, especially for ABR and UBR traffic.

The ATM CBR and VBR service categories have been designed to avoid network congestion conditions when implemented with robust connection admission controls and usage parameter controls. ABR and UBR service categories have been defined to take advantage of the excess capacity beyond that required for CBR and VBR connections.

In the current Internet, the network components are assumed not to do any congestion control. As such, the endpoints rely on end-to-end protocols, such as TCP/IP, to regulate the flow of traffic, based upon packet loss. This model can also apply for TCP/IP over an ATM transport infrastructure.

One of the effective methods of congestion control in high speed ATM networks is to control the user traffic at the user-network interface [Chao 1992, Cooper 1990, Gerla 1990, Hiramatsu 1991, Jain 1990, Katevenis 1987, Khosrow 1991]. This type of control at the user-network interface is also called rate-based access control [Chao 1992].

Currently, the most widely known rate-based access control system is the leaky-bucket scheme in which a user is periodically allocated a certain number of tokens based on the bandwidth allocated for the user at the initiation of service request [Gerla 1990]. If the user exceeds the allocated input rate, the excess traffic from the user is discarded by the network. In a sense, the rate-based access control serves as a traffic shaper at the edge of the network.

The rate-base access control schemes have been extensively studied by the scientific community [Chao 1992, Gerla 1990, Khosrow 1991]. Additionally, Chao [Chao 1992] studied a rate control system based on the shared buffer scheme to regulate outgoing traffic at each intermediate ATM node to prevent congestion.

The ability to function gracefully in an overload situation will be the key benchmark requirement for all ATM switches and will be the critical function most analyzed in the comparison of ATM switches from multiple vendors. An ATM switch architecture must therefore provide robust congestion management support, multiple service classes and per VC accounting with the capability of operating on a cell time basis.

Specifically, a congestion management implementation should include per VC accounting for maximum cells, congested cells, current cells, dropped cells and received cells all processed on a per port, per cell time basis. And finally, an ATM switch which provides support for ABR will have to incorporate additional accounting features which of course must be handled on a cell time basis requiring processing bandwidth and memory.

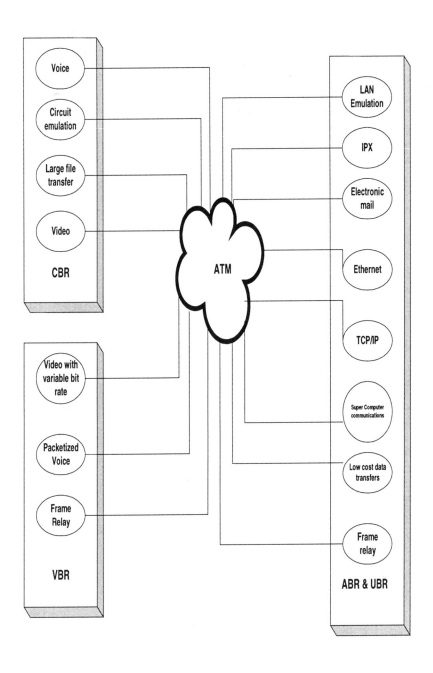

Figure 8-21 The various applications and the service types.

Table 8-7 lists various ATM switch vendors who provide a comprehensive set of congestion control mechanisms to optimize bandwidth utilization while maintaining a high level of service degree during periods of overload.

Table 8-7

Company Switch	Newbridge 36170	Ascend CBX 500	Nortel Concorde
Flow Control Features	EFCI Marking Explicit rate VS/VD	ABR; EFCI & relative rate, Explicit rate VS/VD RM cell priority	EFCI marking Per-VC queuing
Traffic Policing	Yes, Dual Leaky Bucket GCRA	Yes	Yes
CAC	Yes	Yes	Yes

Preventive congestion control is used to avoid congestion and guarantee the Quality of Service (QoS) of the active connections in case network resources run low. The ATM network informs the traffic sources of impending congestion using the ABR feedback mechanisms. On receiving this information, the traffic sources stop increasing their traffic. The objective is to reduce the overall traffic in a way that the network never reaches an undesirable state of congestion.

Reactive congestion control is used to overcome congestion situations while guaranteeing the QoS of the CBR and the real-time and non-real time VBR connections. Again, ABR feedback mechanisms are applied. On receiving a congestion indication the traffic sources immediately reduce their traffic in a standardized manner.

There is a problem with ATM networks carrying packet traffic, since higher layer packets or Protocol Data Units (PDUs) are segmented when they are adapted to the smaller ATM cells. Even partial packet loss requires the re-transmission of the entire packet. In other words, even during periods of minor congestion when only a few cells are actually discarded, there is the possibility that if each discarded cell comes from a different PDU, it can lead to a dramatically escalating congestion situation.

Various mechanisms are developed by ATM switch vendors to deal with this problem. For example, to fix the problem of flooding the network with re-transmitted higher layer PDUs, NewBridge MainStreetXpress products provide Early Packet Discard (EPD) and Partial Packet Discard (PPD). EPD and PPD are applied for ABR and UBR traffic of AAL type 5 connections which are used especially for this packet traffic.

Packet discard means that if ATM cells have to be discarded because of congestion, cells are discarded on a packet level rather than a cell level. This drastically reduces useless traffic caused by the transfer of corrupted data packets which have to be re-transmitted by the sender.

Early Packet Discard   When congestion occurs and buffers are filling, EPD discards new packets arriving at a queue, i.e., all cells associated with a new packet are discarded. The remaining buffer space can then be used for ATM cells belonging to packets that already have entered the queue.

Partial Packet Discard   If EPD does not remove congestion and cells arriving at a queue have to be discarded because of buffer overflow, PPD is applied. PPD discards all subsequent cells associated with the same packet, rather than just a few cells within the packet during buffer overflow.

EPD and PPD complement each other very effectively, improving the chances for more complete packets to get through. EPD maximizes the chances for already queued packets to leave the queue successfully, and PPD minimizes the number of packets becoming invalid in the queue.

Both mechanisms have proven to drastically improve performance for data applications, such as TCP/IP. It effectively ensures that bandwidth is not wasted through the network sending partial frames.

Chapter 9

ATM INTERWORKING STANDARDS

The ATM standards bodies such as ATM Forum and ITU-T developed sets of ATM interworking standards to allow interoperability with various network environments. The main thrust behind this attempt is to gain wider application coverage for ATM as well as to allow ATM to coincide with legacy networks. In this chapter, we will review some of the most significant interworking standards such as LAN Emulation (LANE), Classical IP over ATM (CLIP), Multiple Protocols over ATM (MPOA), and Voice and Telephony over ATM (VTOA).

I. LAN EMULATION

The sheer size of the currently installed legacy LAN networks cannot be overlooked if ATM is to be considered as a major player in the data networking market. Particularly, Ethernet technology among the legacy LAN networking technologies captured wide acceptance in recent years and became a dominant technology to reckon with. At the present time, Ethernet still enjoys wide acceptance and support. There is a strong interest in the data networking industry to enhance Ethernet to keep it viable for data networking technology for the foreseeable future.

ATM technology can not achieve its goal to become a widely accepted data networking standard by ignoring interoperability with the legacy LAN networking technologies such as Ethernet, Token-Ring and FDDI. Its success in the data networking market will mainly be determined by its strength in accommodating these legacy LAN technologies, particularly the Ethernet technology, and not by its robust bandwidth capacity and scalability features.

Realizing the significance of interoperability with the already installed legacy LAN networks, the ATM Forum issued the first version of ATM LAN Emulation (LANE 1.0) standard in 1995. The key objectives of the LANE standard are:

- Interconnectivity between ATM-based LAN networks and legacy LAN networks.
- To allow existing software applications to access ATM network using traditional LAN protocol stacks such as TCP/IP, IPX, NetBIOS, APPN and AppleTalk.

A. LANE Architecture

LANE is based on a client/server architecture. Each Emulated LAN (ELAN) contains many LAN Emulation Clients (LECs) (end-users) and a single LAN Emulation Service (network side). The LAN Emulation Service is composed of a LAN Emulation Server (LES), a LAN Emulation Configuration Server (LECS), and a Broadcast and Unknown Server (BUS). This basic client/server architecture is shown in Figure 9-1.

There are significant differences between the way ATM and legacy LAN networks operate. The legacy LAN networks have evolved around shared medium architectures. The communication between end-stations is typically connectionless. The shared medium is a very natural environment for broadcast and multicast functions. These functions are performed effortlessly in a shared medium. On the other hand, ATM is basically a connection-oriented technology. Broadcast and multicast functions require significant amounts of effort in ATM networks. What LANE actually does is to emulate this shared medium environment for ATM through the LANE services. The LANE service components create a common environment where end-stations receive a broadcast type connectionless service.

ELANs can be overlaid over an ATM network so that each member of an ELAN does not have to be physically located in the same building to be connected to the same LAN segment as is the case with the legacy LAN networks. In addition, an end-user can be part of more than one ELAN group. This kind of flexibility is possible because LANE allows logical grouping over an ATM network independent of the physical locations of the group members.

LANE is capable of providing connectivity between ATM-based end systems and end systems attached to legacy LANs. It is even capable of connecting legacy LAN segments over an ATM network. This capability of LANE is illustrated in Figure 9-2. LANE is also compatible with legacy LAN bridging methods such as Transparent Bridging and Source Routing Bridging.

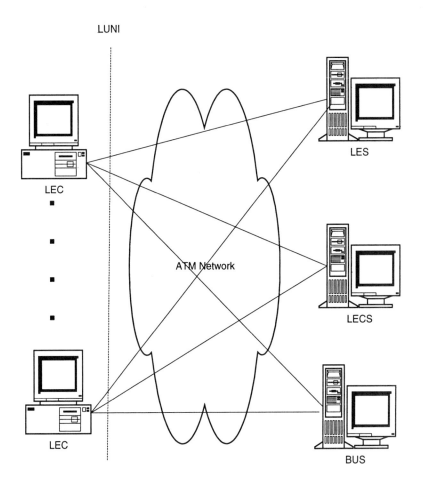

Figure 9-1 Basic LANE architecture.

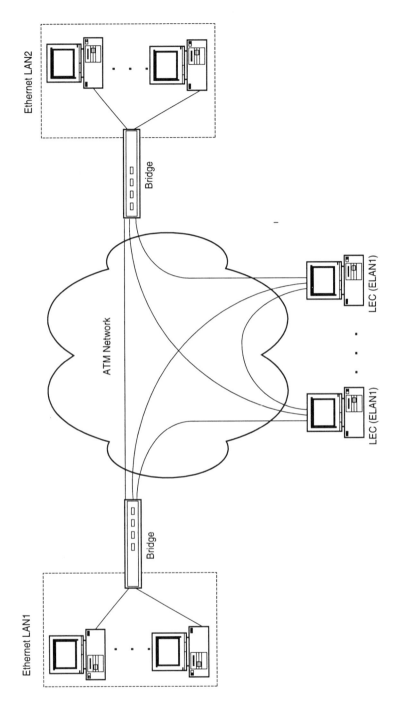

Figure 9-2 LANE interconnectivity.

In its current form, LANE is designed to emulate Ethernet (IEEE 802.3) and Token-Ring (IEEE 802.5) LAN technologies. In addition, LANE only allows single media type emulation within an ELAN. In other words, it does not allow mixing of end-stations running Ethernet and Token-ring emulation within an ELAN. However, it is possible to interconnect two such ELANs through an ATM router which belongs to both ELANs as shown in Figure 9-3.

The network layer protocols like IP and IPX interface to the MAC (Medium Access Control) protocols of legacy LANs through well-defined interface procedures such as Open Data-Link Interface (DDI), Network Driver Interface Specification (NDIS), and Data Link Provider Interface (DLPI). These procedures define how to access a MAC driver via well-defined primitives and parameters. LANE provides in a sense the same type interfaces to access the underlying ATM network. Hence, the main function of LANE is to provide the same type of interface (look and feel) of a legacy LAN to network protocols so that existing LAN applications can run over an ATM network without knowing anything about the network. In other words, LANE hides the ATM network from the upper layers of protocol stack and provides the appearance of a MAC interface. The details of the operation of the ATM network is taking care of by LANE as shown in Figure 9-4.

B. LANE Components

The LEC resides in the end-stations and provides the LUNI interface for the end-stations. The LUNI interface emulates MAC level services for Ethernet or Token-Ring to upper layers in the end-stations. Among the services provided include address resolution, data forwarding and control functions.

The LES is responsible for the control coordination function within an ELAN. It performs a registration function and maintains a table of all registered (active) members of an ELAN. LECs register their MAC addresses they represent with the LES. LES performs the address resolution function for LECs by maintaining a MAC and/or route descriptor to the ATM address conversion table. The LES resolves an address resolution request either by itself if possible (i.e., if the requested address is registered with the LES) or forwards the request to other clients through BUS so they may respond.

The BUS performs two major functions for ELAN: broadcast/multicast and address resolution for unknown clients, also known as the unknown server function. When a LEC wants to broadcast to all members of the ELAN using MAC address "FFFFFFFFFFFF", the broadcast task is performed by the BUS component. Similarly, the multicast is a derivative of the broadcast function in which case the broadcast is targeted to a specific set of members of the ELAN, i.e., to a multicast group. The first unicast data exchange between two LECs has to be done through BUS if the destination LEC has not registered with the LES yet. The BUS server broadcasts the first message so that the unknown LEC will receive and respond, thus providing its ATM address. Once the ATM address of this LEC is known, a direct VCC patch is established between the two LECs.

The BUS component of LANE is responsible for the administrative functions. It manages assignment of LECs to particular ELANs. An LEC can be a part of more than one ELAN. After validating an LEC request to join to a particular ELAN by using policies and a configuration database, the BUS allows this LEC to become an active member of the ELAN by providing the ATM address of its LES server. This logical assignment capability allows independence from the physical location. Hence, LECs from diverse locations can become a part of an ELAN. On the contrary, for the legacy LANs, the membership for the LAN is dictated by the physical location.

C. LAN Emulation User to Network Interface (LUNI)

The LUNI interface between the LE Clients and LE Service provides four basic functions for the operation of LAN Emulation. These four basic functions are summarized in Table 9-1.

Table 9-1 LUNI interface functions.

Initialization	- Obtaining the ATM address(es) of LE Services - Joining and leaving a particular ELAN - Declaring the need for Address Resolution
Registration	- Specifying the list of MAC addresses represented by the LEC - Specifying the list of Source Route descriptors represented by the LEC for Source Route bridging
Address Resolution	- Finding the ATM address of an LEC with particular MAC address
Data Transfer	- Encapsulation of LE-SDU in an AAL5 frame and sending by the LEC - Forwarding the AAL5 frames by LE Service - Receiving and decapsulating the AAL5 frames by LEC

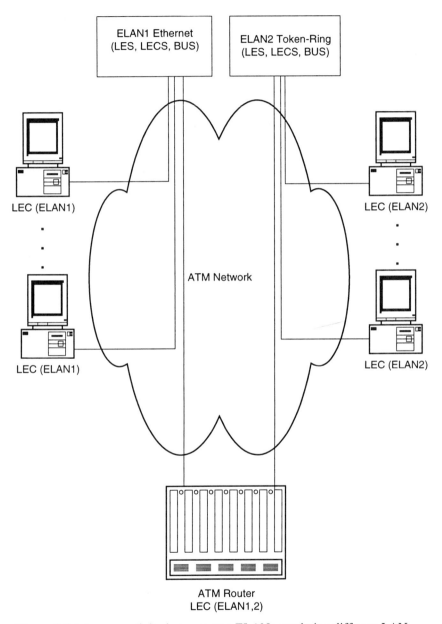

Figure 9-3 Interconnectivity between two ELANs emulating different LAN types (Ethernet and Token-Ring).

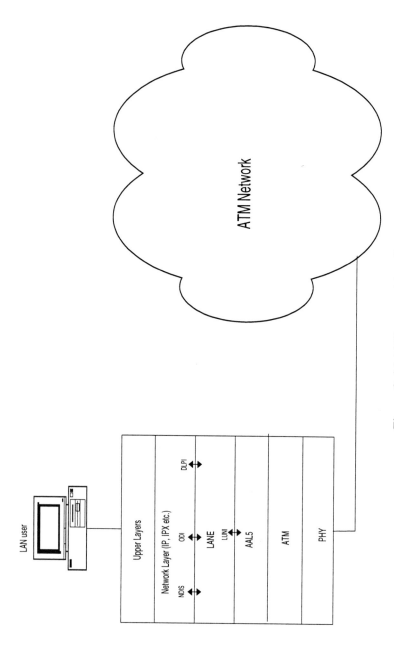

Figure 9-4 LANE protocol interface.

D. LANE Connections

The LANE protocol uses a set of well-defined ATM connections among its components: LE Clients and LE Service element. These connections can be Switched Virtual Channel Connection (SVCC), Permanent Virtual Circuit or a combination of both. While the SVCC type requires ATM signaling function, i.e., call setup and call clear down procedures, the PVCC connection is handled by the layer management.

There are two types of VCC connections required for communication between LE Clients and LE Service components: control and data VCCs. The control VCC is used for carrying control traffic while the data VCC is used for transfer of encapsulated Ethernet/Token-Ring frames. Each control/data VCC is dedicated for only one ELAN. The control VCCs between an LE Client and LE Service components LES and LECS are shown in Figure 9-5. The data VCC can be set up between two LE Clients or between an LE Client and BUS as shown in Figure 9-6.

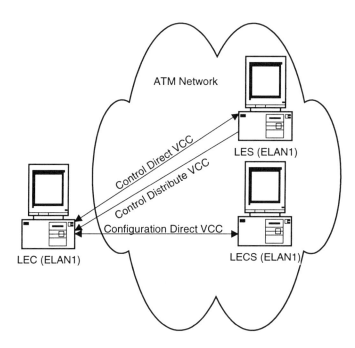

Figure 9-5 LANE control connections.

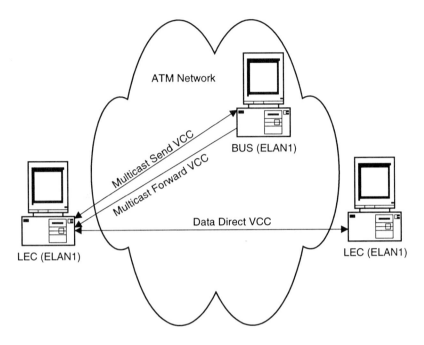

Figure 9-6 LANE data connections.

The control VCCs are established during the LEC initialization sequence strictly for the purpose of carrying LE_ARP traffic and control frames.

The Configuration Direct VCC is a bi-directional point-to-point control connection between LEC and LECS. It is established using B-LLI signaling during the LEC connect phase to obtain configuration information such as the address of LES. The B-LLI signaling is used to indicate that the connection is carrying "LE Control" packet formats. Once the LEC connection to ELAN is established, it is optional to maintain this connection for further configuration inquiries throughout the connection. An LEC can use this connection to inquire about other LECs during its participation in the ELAN.

The Control Direct VCC is a bi-directional point-to-point control connection between LEC and LES. It is established during the LEC initialization phase. In contrast to the Configuration Direct VCC, the LEC and LES are required to preserve this VCC connection throughout the participation of the LEC in the ELAN.

The Control Distribute VCC is a unidirectional point-to-point or point-to-multipoint control connection from LES to LEC(s). This is an optional control connection which the LES may establish during the initialization of an LEC. Once it is established, the LEC and LES are required to maintain this control connection throughout the participation of the LEC in the ELAN.

The Data Direct VCC is a bi-directional point-to-point data connection between LECs to exchange unicast data. When an LE Client wants to send a packet to another LEC for the first time and it does not have the ATM address of this destination LEC, the source LEC issues an LE_ARP request to BUS in order to obtain the required destination ATM address. Upon receiving an LE_ARP response with the ATM address of the destination LEC, the source LEC sets up a Data Direct VCC to the destination LEC to be used for the remainder of the data exchange. If the source LEC is unable to set up this new connection due to lack of resources, it is expected to clear down one of its existing Data Direct VCC connections in order to free up the nessary resources to set up this new connection instead of keep repeating the LE_ARP request to BUS.

The Multicast Send VCC is a bi-directional point-to-point data connection between LEC and BUS. An LEC establishes this connection using the same procedure which was described previously for Data Direct VCCs. When the LEC wants to send multicast data to multiple elements of the ELAN, it uses this VCC to pass the multicast data to the BUS for distribution. The LEC also uses this VCC for sending initial unicast data to an unknown destination through the BUS. The LEC is required to maintain this VCC for the duration of its participation in the ELAN and to accept data from this VCC which the BUS might use to send data or a response back to the LEC.

The Multicast Forward VCC is a unidirectional point-to-point or point-to-multipoint data connection set by the BUS to the LEC during the initialization of the LEC for the ELAN sign up. The primary function of this VCC is to distribute data from the BUS to the member LECs of the ELAN. The LEC is required to accept the connection request and to maintain this VCC for the duration of its participation in the ELAN. The BUS may choose to forward data over either the Multicast Send or Multicast Forward VCC with the assurance that the LEC will not receive duplicate data from both VCCs.

E. Basic Operations Flow for LEC

We next describe the basic operational flow for LE Clients. The key elements of this operational flow include Initialization, Active Operation and

Recovery phase. The operational flow as shown in Figure 9-7 starts with the initial state. The initial state involves the definition of key ELAN parameters such as ELAN name, maximum frame size allowed and essential addresses such as LE Configuration Server, etc. The LEC uses these parameters for the ELAN sign-up process. It is possible that these parameters may be defined for more than one ELAN for the LEC so that the LAN can participate in more than one ELAN particularly if the LEC is to serve as a bridge/router between several ELANs and/or Legacy LANs. It is also possible that these parameters include a predefined LE Server address so that the LEC can directly use this information to discover the LE Service without using the optional LE Configuration protocol. This predefined LE Service approach is somewhat simpler but restrictive in a sense that we lose the flexibility to reassign LE Service on demand.

When an LEC decides to join a particular ELAN, it first goes through the LECS Connect phase. In this phase, the LEC sets up a Configuration Direct VCC to the LECS to obtain the necessary configuration information such as the LES address, etc. Upon establishment of this VCC, the LEC goes into the next phase which is the Configuration phase.

In the Configuration phase, the LEC receives information about the LE Service through its Connection Direct VCC to the LECS and proceeds to the Join phase.

In the Join phase, the LEC establishes its Control Direct and Control Distribute VCCs to the LE Server. Upon successful connection to the LES, the LEC receives information which is critical to its participation in the ELAN. This information includes a unique LEC identifier (LECID), maximum frame size allowed and its LAN type, i.e., Ethernet or Token-Ring. Following the successful completion of Join phase, the LEC precedes with the Initial Registration phase.

In the Initial Registration phase, it is possible for a LEC to register additional MAC addresses and /or Router Descriptors in addition to the single MAC address used in the Joining phase. Through this Initial Registration process, the LEC can validate the uniqueness of its local addresses before becoming operational in order to eliminate the possibility of any address conflict within the ELAN. After this address validation process through the Initial Registration phase, the LEC precedes with the BUS Connection phase.

In the Bus Connect phase, the LEC sends an LE_ARP to the "all ones" broadcast MAC address in order to set up a Multicast Send VCC to the BUS. Upon receiving the LE_ARP response containing the BUS address, it establishes this connection. In turn, the BUS establishes a Multicast Forward

VCC to the LEC. Upon successful completion of the BUS Connect phase, the LEC is considered operational within the ELAN.

The LEC Recovery phase involves returning to the Initial State in all failure cases except the loss of BUS connection in which case the LEC goes back to the BUS Connect phase to re-establish the BUS connections.

F. LANE Connection Management

A LANE environment can be set up to operate with Permanent Virtual Connections (PVCs) or Switched Virtual Connections (SVC) for the connections among the LANE elements (LEC, LES and BUS). When operating with PVC connections, the PVCs are set up manually using the ATM Layer Management entity. On the other hand, if the LANE is operating with SVC connections, the ATM UNI signaling protocol is used to set up the required connections. At least the best effort quality of service is required for these connections.

When using SVC for Data Direct VCC, a slightly modified version of ATM call setup procedure is used to guarantee readiness of both sides (calling and called parties) before the data transfer takes place. This modified call setup protocol is shown in Figure 9-8.

The first part of the call setup involving SETUP, CONNECT and CONNECT_ACK messages is the same as a typical SVC setup. The second part of the procedure involving READY_IND and READY_QUERY is specific to the LANE environment. The main reason for the additional message sequence is to ensure end-to-end validation that both parties are ready to exchange data. For example, it is possible that the called party may receive the CONNECT_ACK message before the calling party gets the CONNECT message. Such an undesirable timing has to do with the fact that the CONNECT_ACK message may be sent from the local switch where the called party is connected. If the CONNECT_ACK message was to be interpreted as a GO signal by the called party, then the called party would possibly start sending data to the calling party before the CONNECT message is received by the calling party. The CONNECT message carries critical information such as the indication of successful allocation of the VPI/VCI numbers for the Data Direct VCC. Hence the calling party cannot make itself ready to receive data from the called party until it receives such confirmation.

189

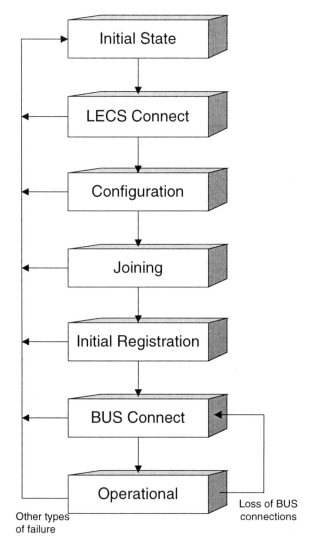

Figure 9-7 Basic LEC operational flow.

The READY_IND and READY_QUERY messages provide end-to-end confirmation to both sides. When the called part receives the CONNECT_ACK message, it starts a timer for the supervision of READY_IND. If the READY_IND is received before the timer expires, the called party stops the timer and goes into ready state to exchange data with the calling party. However, if the timer expires before receiving the READY_IND, then the called party sends a READY_QUERY message to request confirmation from the called party. Both sides can initiate READY_QUERY to confirm the readiness of the other side. Both sides are required to respond to the reception of a READY_QUERY message on an active VCC with a READY_IND message.

II. CLASSICAL IP AND ARP OVER ATM

The ATM LAN Emulation which we discussed previously is designed to provide Layer 2 switching and bridging services to interconnect existing legacy LAN systems such as Ethernet and Token Ring-based LANs to be interconnected through the high-speed ATM campus backbone network. The LANE emulates the MAC layer of the LAN protocol stack and is transparent to network (Layer 3) level protocols such as IP and IPX. In addition to Layer 2 emulation of Ethernet and Token Ring in the LAN environment Classical IP and ARP over the ATM concept has been described in RCF 1577 by the IETF organization in order to extend the use of ATM in the IP-based LAN environment. This protocol is also referred to as Classical IP over ATM (CLIP).

The key objectives of CLIP is to take advantage of higher wire speed offered by ATM technology while protecting investment in today's IP based applications and to help eventual migration of legacy IP-based legacy networks such as LANs to ATM-based networks.

The CLIP retains the IP characteristic at the Layer 3 (network) level for the existing applications while switching Layer 1 and 2 from Ethernet/ Token Ring to ATM. In other words, CLIP models an IP network over an ATM network so that it is possible to run existing IP-based applications without modifications over an ATM network.

As shown in Figure 9-9, CLIP creates a Logical IP Subnetwork (LIS) over an ATM network. It is possible to create many such LIS over the same ATM network. A typical LIS is composed of many members (hosts and routers) and a single ATMARP server. In a typical IP-based LAN network, the subnetwork boundary is defined by the physical location. On the other hand, the LIS provides a logical boundary for the IP subnetwork; thus the location of its members is not a limiting factor.

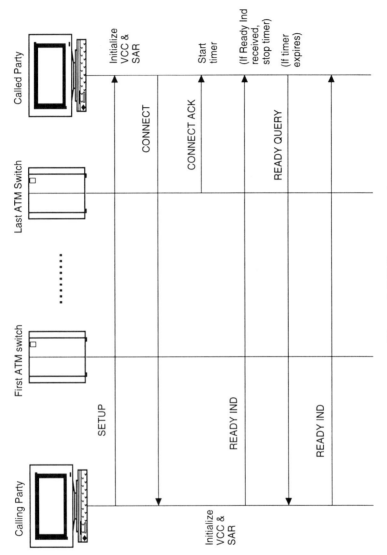

Figure 9-8 Call setup procedure.

A. LIS Configuration and Operation

There are two flavors of LIS configurations: one is based on PVC and the other is based on SVC. In the PVC case each member is connected to an ATMARP server via preestablished PVCs. In the case of using SVC, each member joins the LIS by setting up a SVC connection to the known address of the ATMARP server. Each member as part of its joining process registers with the ATMARP server by providing its IP and ATM address as shown in Figure 9-10.

When a member of an LIS wants to communicate with another IP destination, it requests an IP to ATM address translation for the destination station from the ATMARP server by sending an ATMARP Request. Upon receiving the ATM address of the destination in an ATMARP Response message, it will set up a SVC connection to establish the communication with the destination. This scenario is illustrated in Figure 9-10. If the destination is located in another LIS, the ATMARP server will return the address of a default router which serves as a bridge to the other LIS domain.

As shown in Figure 9-11, several routers are configured as members of multiple LISs to serve as a bridge to carry traffic between these logically separated LISs. Both LANE and CLIP suffer from this same drawback in terms of relying on routers to carry traffic between the logically separated subnetwork domains. This is true even if both source and destination are residing in the ATM network, i.e., it is possible to set up a direct VC (short-cut) between the two stations over the ATM network. Reliance on routers create extra latency since these routers have to check every data frame header to make a routing decision. This reliance also creates bandwidth bottlenecks since the traffic between the subnetwork domains has to be channeled through these routers. The Next Hop Routing Protocol (NHRP) is specifically designed to address these problems.

RCF 1577 defines the key operational requirements for the LIS environment as:

- All members should have the same IP network/subnet number and address mask.

- All members of a LIS should be directly connected to the ATM network and be able to set up direct VCs to each other (fully meshed connectivity).

- The same Maximum Trasmission Unit (MTU) has to be used for all VCs in a LIS. The default MTU is 9180 octets. However, this can go up to the AAL5 limit of 64K-1 octets as long as each member is configured to use the same MTU value.

- The default encapsulation method for IP packets is LLC/SNAP.

The NHRP replaces the traditional routers with the NHRP Servers (NHS). These NHS servers perform the address resolution to enable the source to set up direct VC (short-cut) to a destination in another subnetwork domain.

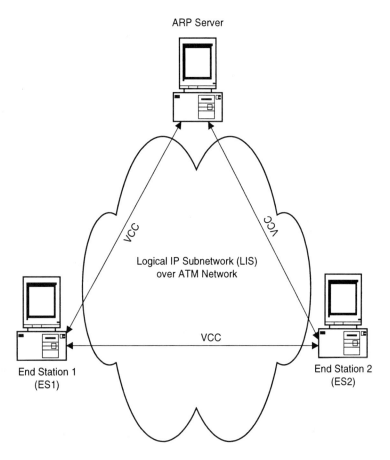

Figure 9-9 Classical IP over ATM configuration.

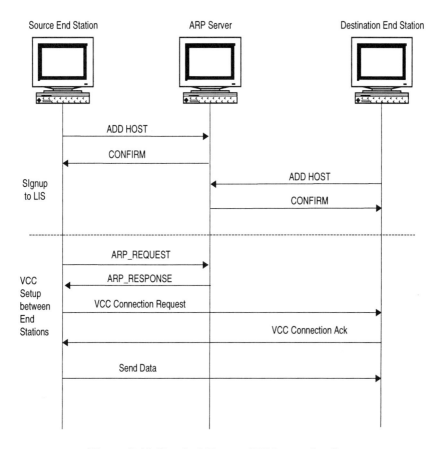

Figure 9-10 Classical IP over ATM operation flow.

III. MULTI-PROTOCOL OVER ATM (MPOA)

It can be safe to say that MPOA is the next stage in the evolution of ATM based LAN networking. It is a result of experience gained from LANE and CLIP. The experience with LANE and CLIP revealed some significant deficiencies in the way the standard bodies approached the problem of transition from legacy LAN to ATM-based LAN environment. One of the key deficiencies of LANE and CLIP was the fact that it was still necessary to use classical routers to carry traffic between the logical subnetworks created by these protocols even if the logical subnetworks resided over the same ATM

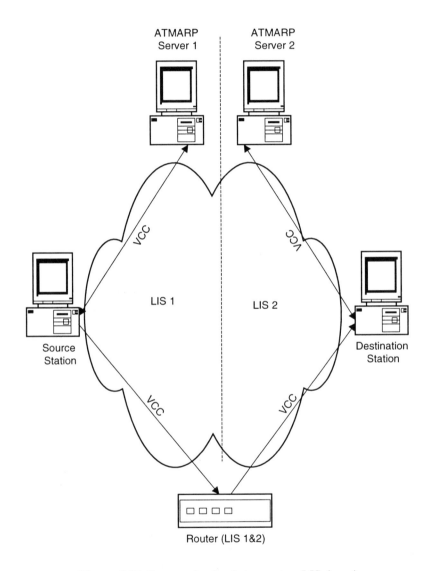

Figure 9-11 Communication between two LIS domains.

network. As mentioned earlier, these routers cause excessive latency and become the bottleneck points, since every packet has to be individually processed by these routers. In addition, since these routers have to regenerate the packets from ATM cells in order to interpret the packet header for routing purposes, this creates additional overhead.

The router concept worked well in the traditional LAN environment when the traffic generated in the LAN segments was IP-based and the amount of traffic that had to go through the routers was a relatively small percentage of the traffic in the LAN environment, i.e., only 20% of the traffic had to go through the router. However, due to a changing traffic pattern, i.e., the reversal of the ratio required higher bandwidth LAN backbone, ATM was introduced to deal with the increased traffic in the LAN backbone. Today, the routers have to have the option to deal with the ATM-based traffic on the LAN backbone. However, the IP based Layer-3 routing requires that these routers have to do a conversion between ATM and IP formats.

The solution provided with MPOA eliminates the bottleneck and overhead created by the classical routers by setting up a VC connection (short-cut) between two end-stations once a traffic flow is detected between them. In other words, the routing function and the physical path are separated from each other to take advantage of the efficiency of ATM-based transport.

MPOA allows seamless networking between ATM and non-ATM subnetworks. The authors of this book believe that MPOA will allow ATM to make significant inroads into the traditional LAN world. The key factor in this progress is the fact that MPOA eliminates the problems associated with the classical routers such as excessive latency and bottleneck for traffic flow by separating switching from routing. The power of MPOA comes from the fact that it not only replaces classical routers but provides significant improvements over them.

MPOA also addresses another problem associated with CLIP which is the lack of multicast capability. It also extends the LANE's multicast capability beyond the ELAN boundaries. MPOA provides multicast capability at Layer-3.

MPOA is a combination of LANE, NHRP, Multicast Address Resolution Server (MARS) and Multicast Server (MCS). MPOA relies on LANE for Layer-2 bridging between ATM-based LAN segments and traditional LAN segments based on Ethernet or Token Ring within an ELAN logical subnetwork. It relies on NHRP for Layer-3 routing between the logical subnetworks.

MPOA also provides a very efficient VLAN platform on an ATM network which may cover large geographical distances.

A. MPOA Architecture

Similar to LANE and CLIP, MPOA also uses the client-server model as shown in Figure 9-12. As we will describe later in this section, the components of MPOA architecture assume a client or server role. An Edge Device or MPOA Host assumes the role of MPOA Client (MPC). A Router assumes a role of a MPOA Server (MPS).

An MPOA Host is an end-station directly attached to the ATM network and contains one or more LANE Clients (LECs) to communicate using LANE services within ELAN domains. An MPOA Host also contains one or more MPOA Clients to communicate across subnet boundaries using Layer-3 networking protocols (i.e., IP protocol).

An Edge Device contains all the functionalties of an MPOA Host. However, it serves as a bridging device between the ATM Network and non-ATM LAN clusters (i.e., Ethernet and Token Ring LAN clusters). Hence, it extends the functionality of MPOA to legacy LANs as part of the ELAN concept.

A Router within the MPOA context contains one or more LECs, one or more MPSs, one or more Next Hop Servers (NHS), zero or more Next Hop Clients (NHC) and zero or more MPCs. Such a router is attached to an ATM network and is a member of one or more ELAN domains in that ATM network.

The main activity between an MPC and MPS is the resolution of a Layer-3 address (i.e., IP address) into an ATM address for a destination in another subnetwork. Once the ATM address of the destination is identified, the MPC sets a shortcut VCC to the destination and forwards the subsequent packets over the shortcut connection, thus bypassing the router path. If the destination resides in the same ELAN then the address resolution takes place at the Layer-2 level by the LANE component of MPOA.

Both MPS and MPC entities maintain caches to reduce overhead associated with the address resolution. Once an entry is made into routing/forwarding tables, this entry is kept for some time for future references. Thus, the subsequent address resolutions are performed through the cached entries locally. Aged entries are periodically removed from the cache tables to make room for new entries.

An MPC observes its outgoing packet traffic to detect flows going to a router with MPS capability which can benefit from a shortcut VCC connection. Once the determination is made, MPC initiates an NHRP-based query to get the ATM address of the destination to set up the shortcut VCC.

For its incoming packet traffic, MPC performs appropriate DLL encapsulation before forwarding them to either a bridge port if it is an edge device or to higher layers of the internal protocol stack if it is a host device.

An MPS is part of a router and it includes its own local NHS and routing functions to respond to MPOA queries for address resolutions from MPCs. An MPS typically converts these MPOA queries into NHRP queries if the address resolution cannot be resolved from the cached entries locally.

B. MPOA Operations

MPOA operations involve 5 basic operations: configuration, discovery, target resolution, connection management and data transfer as summarized in Table 9-2.

The configuration operation involves each MPOA device retrieving its configuration information typically from a LANE LECS server as they become on-line.

The discovery operation allows each MPOA device (MPC and MPS) to recognize other on-line MPOA devices. This discovery process is a dynamic process since the configuration might be changing over time as new MPOA devices become on-line and some on-line MPOA devices become off-line.

The target resolution operation is based on an extended version of NHRP Resolution Request protocol to identify the ATM address of the destination endpoint. The main purpose of the target resolution operation is to switch from the default router-based data transfer path to the ATM-based shortcut connection. The target resolution is triggered by the detection of data flow to a router destination. An Ingress MPC discovers the MAC addresses of MPSs which belong to the same ELAN through the LE_ARP responses it receives from these MPSs. The MPC stores these MAC addresses for flow detection purposes. As it forwards data packets it compares the MAC addresses in the packets against the stored MAC addresses. When a match occurs, the MPC initiates a target resolution operation. A typical target resolution operation flow is shown in Figure 9-11 over the router path.

Data transfer in an MPOA environment occurs either over the default router path or over the shortcut VCC. The degree of shortcut VCCs is the measure of efficient unicast data transfer in an MPOA environment. The shortcut VCCs are established via target resolution or caching mechanism. When a default router path is used, the MPOA edge device behaves like a

Layer-2 bridge. When the shortcut path is used,   the MPOA edge device behaves like a Layer-3 forwarder.

Table 9-2 MPOA operations.

Configuration	- MPCs and MPSs obtain their configuration data from LANE LECS as a default option. Other methods are also possible.
Discovery	- MPCs and MPSs discovering each other through LANE LE_ARP protocol dynamically.
Target Resolution	- Finding the ATM address of an destination with particular networking protocol address (i.e., IP address).
Connection Management	- Creating, maintaining, and terminating control and data VCCs.
Data Transfer	- Forwarding of internetworking data over the default routed path. - Forwarding of internetworking data over the short-cut path.

The connection management is responsible for establishing control and data VCCs as the MPOA components find each other through the discovery operation. These connections are used to carry control information and data flows between the MPOA components. The control flows are very important for the proper operation of MPOA. These control and data flows are shown in Figure 9-13. The control and data flows use LLC/SNAP (RCF 1483) encapsulation as a default option.

The configuration flows are used between the MPOA components (MPCs and MPSs) and the LANE Configuration Server (LECS) to obtain configuration information. The control flows between MPCs and MPSs are used for MPC cache management. The control flows between MPSs (MPS-MPS) are used by Layer-3 routing protocols and NHRP. The control flows between MPCs (MPC-MPC) are used by MPC cache management to remove invalid cache entries. The data flows established over shortcut VCCs between MPCs (MPC-MPC) are used for peer-to-peer data transfer. Finally, the data flows between MPCs and NHSs are used to send unicast data between these entities.

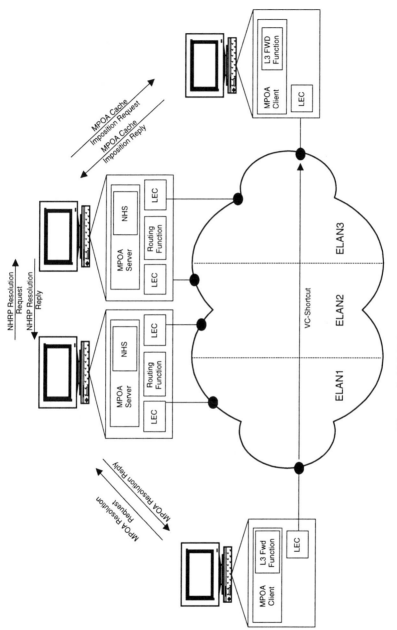

Figure 9-12 MPOA operation.

201

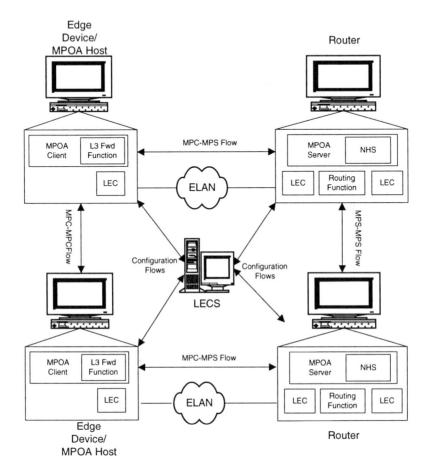

Figure 9-13 MPOA control and data flows.

IV. VOICE AND TELEPHONY OVER ATM (VTOA)

At the present, voice communication still dominates telecommunication activity around the world. In other words, voice communication is still the preferred or only choice for communication. As a recognition of the fact, ATM Forum issued a series of specifications for VTOA application in 1997. The main purpose of these specifications is to allow interworking capability between the ATM-based voice services and the traditional narrowband telephony networks. Each of these specifications addresses a specific aspect of voice and telephony interworking. There are two key aspects of this interworking. The first aspect deals with the end-users (subscribers). The second aspect deals with carrying traditional narrowband voice circuits over ATM networks.

The specification AF-VTOA-0078 describes the Circuit Emulation Service (CES) Interoperability. CES sets the foundation of interworking at the physical layer between ATM and narrowband networks. It describes the protocols to carry PDH circuits (DS1/E1, DS3/E3 and J2) over ATM networks using an AAL1 adaptation layer. These PDH circuit types due to the evolutionary nature of the telecommunication networks have evolved to be used for both voice and data applications. Hence, this specification is applicable to both voice and data applications.

A. Service Types

This specification describes two basic CES types: Structured and Unstructured. It further specifies two sub classes with respect to signaling: Structured with Channel Associated Signaling (CAS) and Structured without CAS. The Structured without CAS is also referred to as a basic structure. These various CES modes are summarized in Figure 9-14.

1. Structured CES

The structured mode is also called Nx64 since a specific physical layer interface DS1, E1, etc. can be divided into one or more logical channels composed of N times 64Kbps time-slots. The range of N is 1-24 for DS1, 1-31 for E1 and 1-96 for J2. Hence, it is possible to carry multiple logical channels over a single physical interface. Each Nx64 channel is carried as a separate VCC across the ATM network. For example, through PBX equipment, several voice channels and data channels can be carried over a

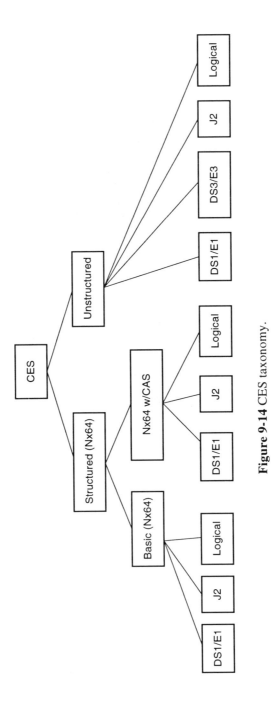

Figure 9-14 CES taxonomy.

single DS1 interface. The Nx64 mode is modeled based on the Fractional T1 (DS1) concept. The logical Nx64 mode describes the CES function without specifying any particular PDH interface. Most recently, there has been an attempt to define structured DS3/E3 in terms of N times T1/E1 by extending the concept of Fractional T1 to Fractional T3 (DS3).

The basic function of the CES Interworking Function (CES-IWF) is to convert PDH time-slots into AAL1 ATM cells at the ingress of the ATM network and do the reverse conversion, i.e., convert AAL1 cells to time-slots at the egress of the ATM network. This interworking operation is shown in Figure 9-15. Within the ATM network, each Nx64 logical channel is set up as a separate VCC. ATM CES interworking offers very flexible mapping of Nx64 channels. For example, multiple Nx64 channels from different physical interfaces can be consolidated into a single physical PDH interface at the egress point.

Another useful aspect of CES is the ability to transfer only the used time-slots of a PDH interface at the ingress point. For example, if a DS1 interface has only 4 of its time-slot assigned for Nx64 traffic, only these 4 time-slots are carried over the ATM network. This is a significant advantage particularly for low usage PDH interfaces since we do not waste any ATM bandwidth for the unused time-slots. This characteristic of CES is illustrated in Figure 9-16. Also note that the mapping of the used time-slots onto egress PDH interface can be different than the input PDH assignments. This flexibility is very useful when consolidating multiple ingress PDH interfaces into one egress PDH interface.

In the Structured CES mode, each Nx64 channel is assigned to a dedicated AAL1 instance and a corresponding VCC. The protocol architecture of Structured CES is illustrated in Figure 9-17. As shown in Figure 9-17, in the ingress direction, the mapping function maps an Nx64 channel from the PDH physical layer time slots to a particular AAL1 instance. In the egress direction, the mapping functions map each AAL1 instance to the designated time slots of the PDH physical layer.

2. Unstructured CES

The unstructured CES mode uses a wire-level emulation. In other words, the ATM CES transports the whole content of the PDH interface transparently without interpreting its format (bit by bit). This type of CES service provides end-to-end transparent transport function through the ATM network. In this case, a fixed amount of ATM network bandwidth is allocated at the line speed of the PDH interface, i.e., 1.5 Mbps for a DS1 connection. The Unstructured CES is similar to leased lines. The specifics such as the frame format, synchronization and alarming are left to the end-user equipment. For

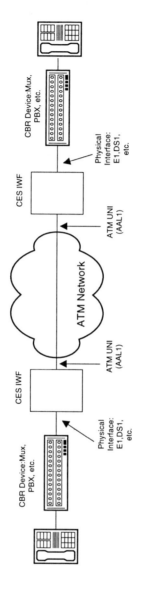

Figure 9-15 A typical voice telephony application over ATM.

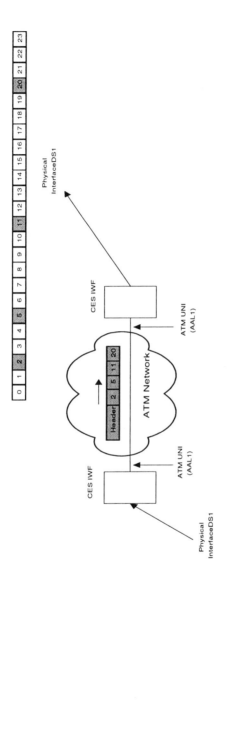

Figure 9-16 Basic principle of ATM CES interworking.

207

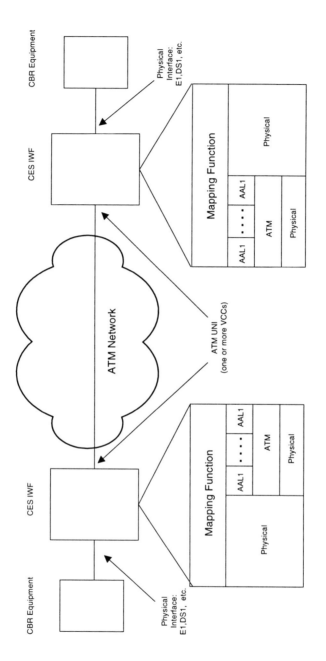

Figure 9-17 Structured CES–IWF protocol architecture.

example, DS1-based end-user equipment may choose to use the standard framing formats such as Super Frame (SF) and Extended Super Frame (ESF) or a non-standard frame format. It may choose to synchronize to the network or run asynchronously with respect to the network. It is up to the end-user equipment to handle alarms and Facility Data Link (FDL). The protocol architecture of unstructured CES is shown in Figure 9-18.

B. AAL1 Structure

The Nx64 operation uses the AAL1 Structured Data Transfer (SDT) mode as described in ITU-T I.363.1. The AAL1 format used for the CES operation is shown in Figure 9-19. When N=1 (i.e., the SDT block carries only one time slot), the AAL1 block pointer is not needed. In this case, the Non-Pointer (Non-P) format with 47-octet payload is used. When N > 1, the Pointer (P) format with 46-octet payload is used. The 1-octet SAR PDU header contains a 4-bit Sequence Number (SN) field and a 4-bit Sequence Number Protection (SNP) field. The SN field includes a 1-bit Convergence Sublayer Indicator (CSI) to indicate the existence of a Convergence Sublayer and a 3-bit Sequence Counter (SC) to make sure that the SAR PDUs are received at a destination in the proper order they are sent from a source. The SNP field contains a 3-bit CRC and 1-bit even parity fields to provide error detection and correction for the SN field.

When operating in the basic structured mode (without CAS signaling), a block is formed by combining the time-slots belonging to this particular Nx64 channel in the order they are received from the physical PDH interface in a single frame. This format is shown in Figure 9-20.

When the Structured with CAS mode is used, the SDT payload is divided into two subsections: the first section contains time slots corresponding to this Nx64 channel from the multiframe PDH data; the second section contains the CAS signaling (A, B, C, and D signaling bits) for these time-slots. This format is illustrated in Figure 9-20.

C. Timing

There are two methods defined for synchronization between source and the destination CBR devices: synchronous and asynchronous clocking modes. In the synchronous clocking mode, both source and destination CBR end-devices attached to the CES-IS entities are synchronized to a reference clock, Primary Reference Source (PRS), from the network through CES-IS entities. In the asynchronous clocking mode, both source and destination end-devices run without synchronizing to the network clock (PRS).

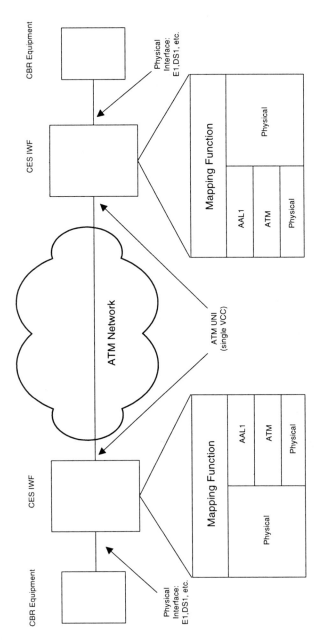

Figure 9-18 Unstructured CES–IWF protocol architecture.

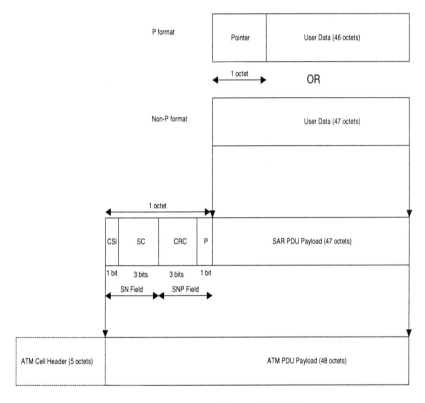

Figure 9-19 AAL1 format for CES.

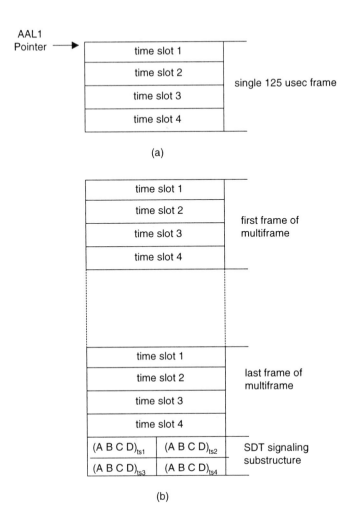

Figure 9-20 AAL1 SDT structure: (a) single frame, (b) multiframe with CAS.

There are two methods used for recovering the source clock in the asynchronous clocking operation: Synchronous Residual Time Stamp (SRTS) and Adaptive Clock Recovery (ACR).

In the SRTS method, it is assumed that both the receiver and sender CES-IS entities are synchronized to a common reference clock from the network. The sender CES-IS entity measures the reference clock cycles with respect to the N service clock cycles. The difference between the service clock and reference clock cycles is then transmitted to the receiver as a 4-bit Residual Time Stamp (RTS) in the CSI bit in SAR-PDU headers corresponding to odd sequence count values of 1, 3, 5 and 7. The receiver CES-IS entity uses the RTS information and the reference clock from the network to synchronize to the service clock of the sender.

In the ACR method, the receiver uses a buffer to synchronize to the service clock of the sender. The receiver writes the received data into the buffer and then reads the data from the buffer using a local clock. The fill level of the buffer is used to control the local clock. The local clock control mechanism tries to maintain the fill level at around the medium position. The ACR method is typically used when the common reference clock (PRS) is not available.

D. ATM Trunking for Narrowband Telephony Services

As part of the VTOA implementation, the ATM Forum described a specification (AF-VTOA-0089) for ATM trunking using AAL1 for narrowband telephony services. It is possible to connect two narrowband telephony networks through ATM trunking as shown in Figure 9-21. This specification addresses IWF issues between the ATM network and narrowband networks in terms of signaling and transport of 64 kbit/s narrowband channels.

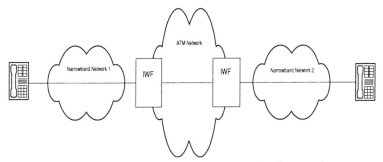

Figure 9-21 ATM trunking for narrowband networks.

In terms of signaling, ATM trunking IWF supports Digital Subscriber Signaling System Number 1 (DSS1) for narrowband ISDN, Private Subscriber Signaling System Number 1 (PSS1) for PBX signaling and Channel Associated signaling (CAS) methods. The CAS signaling described here uses E&M (CAS) with DTMF signaling. The ATM trunking IWF uses Digital Subscriber Signaling System Number 2 (DSS2) for broadband ISDN between two IWF entities connected over an ATM network. The ATM trunking IWF assumes that the same signaling method is used at both ends of a trunk connection. In other words, no signaling protocol conversion is allowed between the two narrowband networks connected through ATM trunks.

It is assumed that the narrowband networks are connected to the ATM network through E1 or DS1 physical interfaces carrying 64 kbit/s channels. These 64 kbit/s channels may carry speech and other voice-band services such as fax and modem services. The ATM trunking IWF uses AAL1 to carry the 64 kbit/s channel content through the ATM network.

The IWF for ATM trunking supports two types of mapping for 64 kbit/s narrowband channels. It can either map each 64 kbit/s narrowband channel to a separate VCC (one-to-one mapping) or map multiple 64 kbit/s narrowband channels to a single VCC (many-to-one). It also uses a separate VCC for IWF to IWF signaling using DDS2 or Network Management (NM). This is illustrated in Figure 9-22.

E. VTOA to the Desktop

The ATM Forum's specification AF-VTOA-0083 describes the interworking function (IWF) between an ATM desktop (Broadband Terminal Equipment, B-TE) connected to an ATM network (B-ISDN network) and Narrowband-ISDN (N-ISDN) network to access N-ISDN voice and other telephony services.

This specification limits the interworking function to handling a single 64 kbit/s voiceband channel between the ATM desktop and N-ISDN network. If the IWF is interfacing to a Private N-ISDN network, then Q.SIG signaling protocol is used. If the IWF is interfacing to a Public N-ISDN network, then ISDN PRI or BRI signaling is used. This is illustrated in Figure 9-23.

For the transport of 64 kbit/s voiceband channel content, either AAL1 or AAL5 can be used by the B-TE equipment. A standard compliant B-TE equipment should be able to accept an incoming call regardless of the fact that it is using either AAL1 or AAL5.

generator with large periodicity is essential for an ATM cell traffic simulator. Additionally, the uniform distribution of the random numbers generated by the random number generator is also essential for unbiased simulation results.

The ATM traffic simulator described in this chapter was developed on an IBM-PC platform with Windows'95 operating system. The simulator was written in Microsoft Visual-C (version 1.0). The Microsoft Visual-C compiler is a 16-bit compiler rather than a 32-bit compiler. Therefore, the random number generator function RAND48 (48-bit random number generator) from a SUN Sparc 4 workstation environment (a UNIX platform) was transformed to a 16-bit version for the Microsoft Visual-C compiler. The simulator was compiled and run on both IBM-PC and Sun workstations. The results from both environments were compared for consistency. The simulator with a 16-bit version of RAND48 on IBM-PC produced the same results as with the Sun workstation version.

The RAND48 is a mixed Linear-Congruential Generator based on the following formula:

$$x_n = (314159269 \, x_{n-1} + 453806245) \bmod (2^{31} - 1) \quad (10.1)$$

The following performance measurements were used for the performance analysis: cell routing delay, throughput, and cell loss probability.

I. QUEUING MODEL FOR THE ATM TRAFFIC SIMULATION

We describe next the queuing model used for the ATM traffic simulation. As mentioned earlier, the ATM switch under consideration is based on the input queuing model. As depicted in Figure 10-3, each input port has its own dedicated queue. Since we use shared buffer space, the input queues are formed as logical queues. The queuing model can be described as a multiserver loss model since each input port is a server and cell loss occurs due to limited (finite) buffer space. The model can be best represented as an G/D/m/B/K using the Kendall's notation described in [Jain 1991], where we assume general arrival distribution (G), deterministic service time (D), multiple servers (m), limited buffer capacity (B) and limited (finite) customers (K). More detailed coverage of the queuing models and their applications to computer and communication systems can be found in [Cooper 1981] and [Tanner 1995]. We next elaborate on each aspect of the queuing model which we used to describe our ATM traffic model.

In an ATM network environment, cells arriving at an ATM node do not necessarily represent a Poisson behavior. As explained in the next section, our ATM traffic model abstracts incoming traffic at three different levels: call level, message level and cell level. Although calls and messages can exhibit memoryless Poisson behavior, cells come in burst mode. This is due to fact that messages are broken into fixed length ATM cells at the ATM network interface. The cells that constitute a user message enter into the ATM network in bursts. That is, there is no separation between the cells belonging to the same burst. Thus, at the cell level there is a statistical correlation between the consecutive cells [Leduc 1994]. Because of this statistical correlation we can not assume Poisson arrival behavior at the cell level. An extensive elaboration on the statistical correlation between the consecutive ATM cells and its influence on ATM traffic modeling can be found in [McDysan 1995] and [Leduc 1994].

It is also possible to consider cell bursts as a unit of arrival event and thus, assume that these cell bursts follow a Poisson arrival behavior. The decision to observe the incoming ATM traffic at the cell level or at the burst level and model the ATM queue accordingly is determined by the characteristics of the server. If the server under consideration processes the incoming cell traffic one cell at a time, i.e., the service time is equal to single cell transmission time (one time-slot), then the observation of the arrival process should be at the cell level. On the other hand, if the server processes cell bursts as a single unit, then the observation of the arrival process can be done at the burst level. In this case, the service time would be variable and can be modeled as an exponential service behavior. At the burst level, the ATM queuing system can be modeled as an M/M/m/B/K queue. In our case, since we are concerned with an ATM switch which processes a single cell as a processing unit, i.e., the service time is constant, the incoming ATM traffic is observed at the cell level and the incoming cell traffic is described as a general arrival process (G). Due to constant service time, the service process is considered to be deterministic (D). Therefore, we define our ATM queue as an G/D/m/B/K queue.

Our queuing model is a multiserver (m) model since we consider each input port as a server with dedicated queues. As well known from the queuing theory, performance of a multiserver queue with a single common queue is much better than the queuing model with a multiserver with dedicated queues for each server. In our case, servers are not generic servers, i.e., an incoming cell has to be handled by a specific input port since the physical transmission path directly terminates to that input port. Of course, it is possible to channel all incoming cells to a single central server which then transmits them to the corresponding output ports. However, in this case the central server has to run N times faster than the single port speed in order to be able to deliver N cells to N output ports in a single time slot for an NxN switch. Given the high-speed operation of a single input port in the order of 100 Mbps, the central server has to run at a very high speed (Nx100 Mbps).

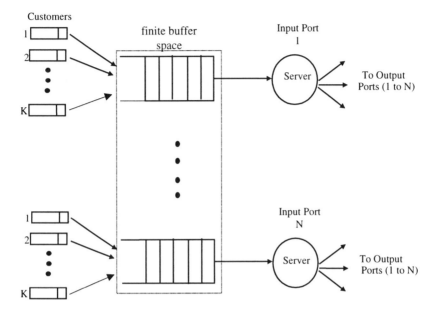

Figure 10-3 Queuing model for the ATM traffic simulation.

The limited buffer capacity (B) is a practical necessity. In our model, a fixed shared buffer space, which is determined at the design stage for an ATM switch under consideration, is dynamically allocated to individual input queues as depicted in Figure 10-3. Unlimited buffer capacity, although feasible for a theoretical model, is not practical to provide unlimited buffer capacity at each ATM switch element. Even if we provide a large buffer space in a real ATM switch, the large buffer capacity tends to cause longer delays by allowing large queue build-ups. In an ATM environment, if an ATM source exceeds its negotiated bandwidth allocation for a long period of time it will have an adverse effect on other calls by creating excessive delays at the ATM nodes along its virtual path. Therefore, the buffer capacity should be chosen according to transmission capacity of an ATM switch which the buffer is suppose to serve.

Since virtual connections are set up in advance during call setup time, ATM cells typically come from a limited number of sources (customers at the cell traffic level). Therefore, we assume a limited customer population (K) for the queuing model. The size of the customer population can vary depending on where the observation is made. At the edge nodes where customers are connected to an ATM network, the population size is fairly small for an input port whereas at the intermediate nodes population size tends to be larger. For the simulation, we consider the ATM traffic at the edge nodes where subscribers are connected. Although the ATM traffic simulator is capable of generating simultaneous calls to different destinations, we are interested in a worst case scenario in which cells arriving at an input port exhibit a high degree of statistical correlation. When cells arriving at the input port of an intermediate ATM node are observed, we expect a higher mixture of cells with different destination addresses. In other words, there is less statistical correlation between the consecutive cells. In this case, it is possible to assume a Poisson arrival process at the intermediate ATM nodes which would certainly provide a better switching performance. However, since we are interested in improving switching performance for the worst case scenario, we assume that each input port is associated with a single subscriber and the subscribers are not allowed to initiate simultaneous calls to different destinations. In the next section we describe the traffic model for the ATM traffic simulator and explain this subscriber behavior in more detail.

II. ATM TRAFFIC MODEL

The traffic model presented here describes the cell generation behavior of the ATM traffic simulator. It reflects the behavior of a typical ATM subscriber. An ATM subscriber starts a call and during the active call state it generates many messages which are broken into fixed length ATM cells. These messages appear as a burst of cells to the ATM switch. Following a quiet period (message interarrival time), the next message is transmitted as another burst of cells. This behavior repeats itself until the call is completed. Then, the subscriber goes through a silent period. At the end of the silent period, the subscriber initiates another call. Due to the memoryless property of interarrival times between consecutive bursts, i.e., the interarrival time between the kth and the $(k+1)^{th}$ arrrivals is independent of the previous interarrival time between the $(k-1)^{th}$ and the kth arrivals, the behavior of an active call can be represented as a two-state (ON/OFF) Markov Process. Figure 10-4 illustrates the corresponding ON/OFF Markov process transition diagram for an active call period in which the subscriber goes through ON-OFF state transitions with certain probabilities. When it is in the OFF state, it remains in this state with the probability $1-\alpha D$ or switches to ON state with the probability αD. Similarly, when it is in the ON state, it remains in this state with the probability $1-\beta D$ and generates cells for the current message. The subscriber switches to the OFF state with the probability βD. This traffic model falls into the category of Markov Modulated Bernoulli Process with two states, MMBP(2). The model provides a good approximation for voice and computer data traffic and a rough approximation for video traffic [Leduc 1994].

According to the model described above, a subscriber initiates a burst with the probability αD which has an exponential distribution and ends a burst with the probability of βD which has a geometric distribution in a particular cell time. The D represents the cell duration in seconds. The average burst duration d (in cells) can be calculated using the standard geometric series as described in [McDysan 1995]:

$$d = \frac{1}{\beta D} \qquad (10.2)$$

The traffic model also defines the behavior of the subscriber at the call level. At the call level, calls are generated via a Poisson arrival process and the silence period between consecutive calls is exponentially distributed. Once a call is active we switch to the cell level traffic model which we described above.

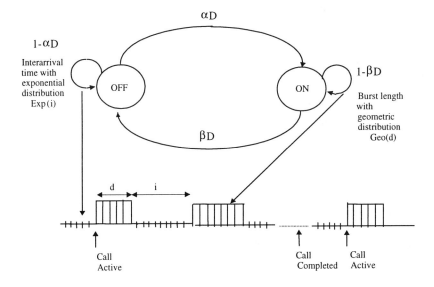

Figure 10-4 Traffic model for the ATM traffic simulator.

223

III. VALIDATION OF SIMULATION RESULTS

When performance of a system is analyzed via a simulation technique, it is necessary to validate simulation results against some reference data. In simulations of complex queuing systems such as Multiserver Loss Models, one common approach for validation is to approximate the queuing system under study into one of the basic queuing models such as M/M/1, M/G/1 or M/D/1 for which analytical models are well defined. The M/M/1 model describes a queue with Poisson arrival, exponential service time, single server, infinite buffer capacity, infinite customer population, and First-Come, First-Served (FCFS) service discipline. The M/G/1 and M/D/1 models differ from the M/M/1 model with respect to the service time. The M/G/1 assumes that the service time follows a General distribution where as the M/D/1 model assumes a Deterministic (constant) service time. The detailed description and mathematical analysis of these basic queuing models can be found in [Cooper 1981, Jain 1991, Tanner 1995]. The operational behavior of the system under consideration is approximated as one of the basic queuing models. Then, the simulation is run for the approximated queuing model. The result of the simulation is compared against the data obtained from the analytical reference model. If the simulation data is consistent with the analytical data, it can be assumed that the simulation environment is reliable.

The ATM queuing model which is used for the simulation can be best approximated as the M/D/1 queuing model since the service time is constant due to fixed cell length. The multiple servers are reduced to a single server. Sufficient buffer space is provided to approximate the infinite buffer size. Additionally, the customer population is also specified to be sufficiently large to approximate the infinite customer population. During the validation runs, we have achieved results very close to analytical M/D/1 data with only 25 ATM customers. Since the message interarrival has exponential distribution by reducing the message length to one cell, the cell arrival process becomes Poisson. Each call generates a single message with the length of one cell. This approximation makes the cell arrival process the M-class arrival process. Since we reduced multiple servers to a single server, then the reduced model becomes an M/D/1 system.

For the validation, the following performance characteristics were used: Mean Queue Length (L_Q), Mean Time-in-System (T_Q/T_S) in service time units (cell units). The Queue Length corresponds to an average number of cells in the queuing system and is given by the well-known Pollacek-Khinchine formula for M/D/1 model [Cooper 1981, Tanner 1995]:

$$L_Q = \rho + \frac{\rho^2}{2(1-\rho)} \tag{10.3}$$

The Mean Time-in-System (T_Q/T_S) for the M/D/1 model in service time units is given by

$$T_Q / T_S = 1 + \frac{\rho}{2(1-\rho)} \tag{10.4}$$

where T_Q is the Mean Time-in-System, T_S is the Mean Service Time, and ρ is the server utilization. The server utilization ρ represents the degree of server utilization and it can be calculated as described in [Tanner 1995]:

$$\rho = \lambda T_S \tag{10.5}$$

where λ is the mean arrival rate. For the M/D/1 queue model, the server utilization is the same as the offered load a.

The comparison of the simulation and the analytical M/D/1 model was done using the performance measures which we just described above. Figure 10-5 illustrates the comparison for the Mean Queue Length while the comparison for the Mean Time-in-System is shown in Figure 10-6. These two comparisons clearly demonstrate that our simulation environment is able to approximate the analytical data very closely and that the ATM traffic simulator is functioning properly.

IV. SIMULATION RESULTS

The single M/D/1 queue model which has been described in Section 10.4 is used as the reference model. The single M/D/1 queue model is defined by assigning all incoming ATM cell traffic to a single input-output port pair. The incoming cell traffic is queued to a single queue and served by a single server (see Figure 10-3 for more detail on the queuing model for the ATM switch). Figure 10-7 shows the influence of statistical correlation between arriving cells on cell delay. The curve for message length m=1 in Figure 10-7 corresponds to a true M/D/1 queuing model since each cell is statistically independent from each other due to the fact that interarrival time between two consecutive messages is a Poisson distribution. Since each message contains a single cell, consecutive cells are statistically independent from each other and follow the same Poisson distribution. Hence, as shown in Figure 10-7, as the message length is increased, the waiting time (cell delay) experienced

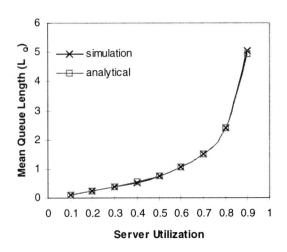

Figure 10-5 Mean Queue Length for M/D/1 model: simulation vs. analytical.

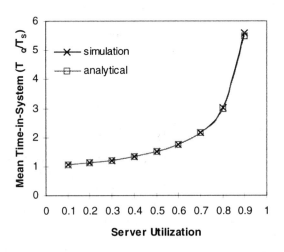

Figure 10-6 Mean Time-in-System for M/D/1 model: simulation vs. analytical.

by the individual cells rises. Similarly, as shown in Figures 10-9 and 10-10, Mean Queue Length and Mean Time-in-System also rise as the message length is increased.

Since the M/D1 queuing model assumes infinite buffer space, a single M/D/1 queue configuration does not experience any cell loss due to lack of buffer space. Every arriving cell is buffered in the queue and eventually processed by the server. By allocating sufficiently large buffer space, i.e., 100 buffers for an 8x8 ATM switch running at 100 Mbps, the infinite buffer space requirement for the M/D/1 queuing model has been satisfied. Since no cell is lost due to lack of buffer space, the increase in message length has no influence on the throughput. This is clearly shown in Figure 10-8.

The single M/D/1 queuing configuration is expanded to multi-M/D/1 queue configuration by assigning a M/D/1 queue for each input-output port pair of an ATM switch. In this configuration each user is assigned to a particular input-output port pair. This configuration is very similar to the Permanent Virtual Circuit (PVC) concept used in ATM networks. The computer simulation results for this configuration are presented in Figures 10-11 through 10-14. The performance measurements for this configuration are in agreement with the single M/D/1 configuration.

The effects of limited buffer size on switch performance are discussed next. The limited buffer size can significantly influence the performance of the switch in terms of cell loss, throughput and cell delay. In the ATM switching environment, simply increasing the buffer size does not always yield optimal performance. Although an increase in buffer size lowers the cell loss rate, it also increases the cell delay at the same time. This is clearly evident in Figures 10-15, 10-16 and 10-17. Therefore, a proper balance must be established between cell delay, cell loss and throughput performance objectives when determining buffer size for an ATM switch.

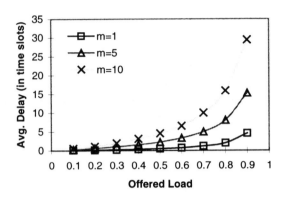

Figure 10-7 Cell delay for single M/D/1 queue configuration: fixed message length (m) and fixed destination address case.

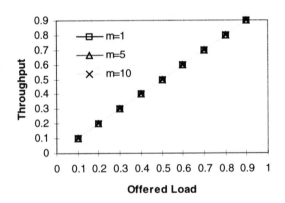

Figure 10-8 Throughput for single M/D/1 queue configuration: fixed message length (m) and fixed destination address case.

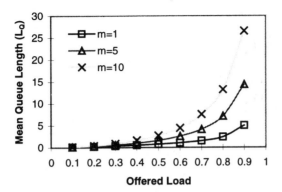

Figure 10-9 Mean Queue Length for single M/D/1 queue configuration: fixed message length (m) and fixed destination address case.

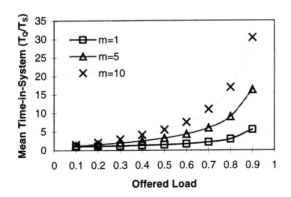

Figure 10-10 Mean Time-in-System for single M/D/1 queue configuration: fixed message length (m) and fixed destination address case.

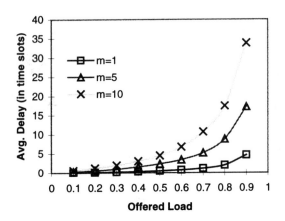

Figure 10-11 Cell delay for multiple M/D/1 queue configuration: fixed message length (m) and fixed destination address case.

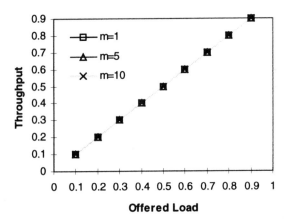

Figure 10-12 Throughput for multiple M/D/1 queue configuration: fixed message length (m) and fixed destination address case.

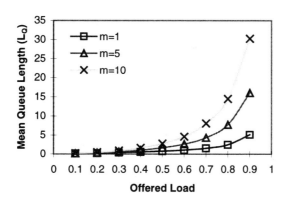

Figure 10-13 Mean Queue Length for multiple M/D/1 queue configuration: fixed message length (m) and fixed destination address case.

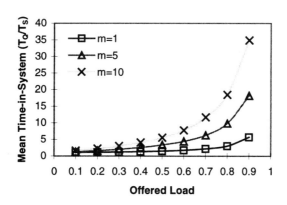

Figure 10-14 Mean Time-in-System for multiple M/D/1 queue configuration: fixed message length (m) and fixed destination address case.

231

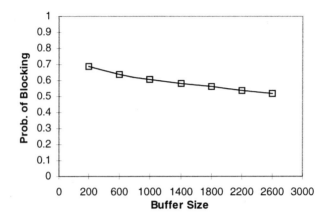

Figure 10-15 Blocking vs. buffer size: offered load = 0.9, switch size = 16x16, speed = 106 Mbs per port.

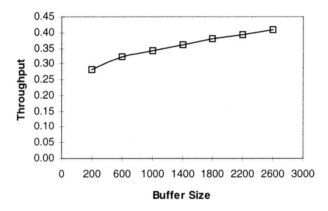

Figure 10-16 Throughput vs. buffer size: offered load = 0.9, switch size = 16x16, speed = 106 Mbs per port.

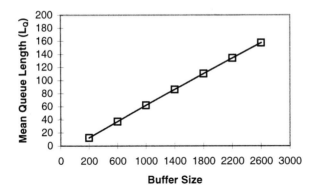

Figure 10-17 Mean Queue Length vs. buffer size: offered load = 0.9, switch size = 16x16, speed = 106 Mbs per port.

Chapter 11

ONGOING ATM STANDARDIZATION ACTIVITIES

I. CURRENT STANDARDS

A. ITU-T Recommendations for ATM

This section provides a list of ITU-T ATM related standardization documents currently in force.

E-series - Overall network operation

•E.177, B-ISDN Routing
•E.191, B-ISDN Numbering and Addressing
•E.716, User demand modeling in Broadband-ISDN

F-series - Telecommunications services other than telephone (operation, quality of service, service definitions, and human factors)

•F.811, Broadband connection-oriented bearer service
•F.812, Broadband connectionless data bearer service
•F.813, Virtual path service for reserved and permanent communications

G-series - Transmission systems and media, digital systems and networks

•G.176, Planning guidelines for the integration of ATM technology into the PSTN
•G.702, Digital Hierarchical Bit Rates
•G.703, Physical/electrical characteristics of hierarchical digital interfaces
•G.704, Synchronous frame structures used at 1544, 6312, 2048, 8488 and 44736 kbit/s hierarchical levels
•G.707, Network node interface for the synchronous digital hierarchy (SDH)
•G.708 (1993) (replaced by G.707)
•G.804, ATM Cell mapping into plesiochronous digital hierarchy (PDH)
•G.805, Generic functional architecture of transport networks
•G.806, (now referred to as I.326)

H-series - Characteristics of transmission channels used for other than telephone purposes

- •H.222.1, Multimedia multiplex and synchronization for AV communications in ATM environment
- •H.245, Control protocol for multimedia communication
- •H.310, Broadband and AV communication systems and terminals
- •H.321, Adaptation of H.320 visual telephone terminals to B-ISDN environments
- •H.323, Visual telephone systems/equipment for local area networks that provide a non-guaranteed quality of service

I-series - Integrated Services Digital Networks

- •I.113, Vocabulary of terms for broadband aspects of ISDN
- •I.121, Broadband aspects of ISDN
- •I.150, B-ISDN ATM functional characteristics
- •I.211, B-ISDN service aspects
- •I.221, Common specific characteristics of services
- •I.311, B-ISDN general network aspects
- •I.321, B-ISDN protocol reference model and its application
- •I.324, ISDN Network Architecture
- •I.326, Functional architecture of transport networks based on ATM
- •I.327, B-ISDN functional architecture
- •I.356, B-ISDN ATM layer cell transfer performance
- •I.357, B-ISDN semi-permanent connection availability
- •I.361, B-ISDN ATM layer specification
- •I.362, B-ISDN ATM adaptation layer (AAL) functional description
- •I.363, B-ISDN ATM adaptation layer (AAL) specification
- •I.363.1, B-ISDN AAL, type 1 specification
- •I.363.3, B-ISDN AAL, type 3/4 specification
- •I.363.5, B-ISDN AAL, type 5 specification
- •I.364, Support of broadband connectionless data service on B-ISDN
- •I.365.1, Frame relaying service specific convergence sublayer (FR-SSCS)
- •I.365.2, Service specific coordination function to provide CONS
- •I.365.3, Service specific coordination function to provide COTS
- •I.365.4, SSCF for HDLC applications
- •I.371, Traffic control and congestion control in B-ISDN
- •I.413, B-ISDN User-Network Interface (UNI)
- •I.414, Overview of recommendations on layer-1 for ISDN and B-ISDN customer accesses
- •I.432, B-ISDN UNI: physical layer specification
- •I.432.1, B-ISDN UNI: physical layer specification - General characteristics

- I.432.2, B-ISDN UNI: physical layer specification for 155,520 and 622,080 kbit/s
- I.432.3, B-ISDN UNI: physical layer specification for 1 544 and 2 048 kbit/s
- I.432.4, B-ISDN UNI: physical layer specification for 51 840 kbit/s
- I.580, General arrangements for interworking between B-ISDN and 64 kbit/s-based ISDN
- I.610, B-ISDN operation and maintenance principles and functions
- I.731, Types and general characteristics of ATM equipment
- I.732, Functional characteristics of ATM equipment
- I.751, ATM management of the network element view

J-series - Transmission of Sound Program and Television Signals

- J.82, Transport of MPEG-2 constant bit rate television signals in B-ISDN

M-series - Maintenance: transmission systems, telephone circuits, telegraphy, facsimile, leased cicuits

- M.3010, Principles of Telecommunications Management Network (TMN)
- M.3611, Test management of the B-ISDN ATM layer using the TMN

Q-series - Switching and signaling

- Q.2010, B-ISDN overview - Signaling capability set 1, release 1
- Q.2100, B-ISDN signaling ATM adaptation layer (SAAL) overview description
- Q.2110, B-ISDN AAL - Service specific connection oriented protocol (SSCOP)
- Q.2119, B-ISDN AAL protocols - Convergence function for SSCOP above the frame relay core service
- Q.2120, B-ISDN meta-signaling protocol
- Q.2130, B-ISDN signalling ATM adaptation layer - Service specific coordination function for support of signaling at the user-network interface (SSCF at UNI)
- Q.2140, B-ISDN ATM adaptation layer - Service specific coordination function for signaling at the network node interface (SSCF at NNI)
- Q.2144, B-ISDN signaling ATM adaptation layer (SAAL) – Layer management for the SAAL at the network node interface (NNI)
- Q.2210, B-ISDN Message Transfer Part level 3 functions and messages using the services of ITU-T recommendation Q.2140
- Q.2610, B-ISDN usage of cause and location in B-ISDN user part and DSS 2

- •Q.2650, B-ISDN interworking between SS7 B-ISDN User Part (B-ISUP) and DSS2
- •Q.2660, B-ISDN interworking between SS7 B-ISDN User Past (B-ISUP) and narrowband ISDN User Part (N-ISUP)
- •Q.2721, B-ISDN User Part - Overview of the B-ISDN NNI Signaling Capability Set 2, Step 1
- •Q.2722, B-ISDN User Part - NNI specification for point-to-multipoint call/connection control
- •Q.2723 through 2727, B-ISDN Extensions to SS7 B-ISDN User Part (B-ISUP)
- •Q.2730, B-ISDN SS7 B-ISDN User Part (B-ISUP) - Supplementary services
- •Q.2761, B-ISDN SS7 B-ISUP - Functional description
- •Q.2762, B-ISDN SS7 B-ISUP - General functions of messages and signals/r
- •Q.2763, B-ISDN SS7 B-ISUP - Formats and codes
- •Q.2764, B-ISDN SS7 B-ISUP - Basic call procedures
- •Q.2931, B-ISDN DSS2 User-network interface (UNI) layer-3 specification for basic call/connection control
- •Q.2932, B-ISDN DSS2 Generic functional protocol - Core functions
- •Q.2933, B-ISDN DSS2 Signaling specification for frame relay service
- •Q.2951, Stage 3 description for number ID supplementary services using DSS 2
- •Q.2955, Stage 3 description for community of interest supplementary services using DSS 2
- •Q.2957, Stage 3 description for additional information transfer supplementary services using DSS 2
- •Q.2959, B-ISDN DSS2 Call Priority
- •Q.2961, B-ISDN DSS2 Support of additional parameters
- •Q.2962, B-ISDN DSS2 Connection characteristics negotiation during call/connection establishment phase
- •Q.2963, B-ISDN DSS2 Connection modification - Peak cell rate
- •Q.2964, B-ISDN DSS2 Basic look-ahead
- •Q.2971, B-ISDN DSS2 UNI layer 3 specification for point-to-multipoint call/connection control

B. ATM Forum Specifications

Below is a listing of all specifications completed and approved by the ATM Forum since its inception in 1991.

B-ICI

•B-ICI 1.0	af-bici-0013.000	Sep, 1993
•B-ICI 1.1	af-bici-0013.001	
•B-ICI 2.0 (delta spec to B-ICI 1.1)	af-bici-0013.002	Dec, 1995

| •B-ICI 2.0 (integrated specification) | af-bici-0013.003 | Dec, 1995 |
| •B-ICI 2.0 Addendum or 2.1 | af-bici-0068.000 | Nov, 1996 |

Data Exchange Interface

| •Data Exchange Interface version 1.0 | af-dxi-0014.000 | Aug, 1993 |

ILMI (Integrated Local Mgmt. Interface)

| •ILMI 4.0 | af-ilmi-0065.000 | Sep, 1996 |

Lan Emulation/MPOA

•LAN Emulation over ATM 1.0	af-lane-0021.000	Jan, 1995
•LAN Emulation Client Management Specification	af-lane-0038.000	Sep, 1995
•LANE 1.0 Addendum	af-lane-0050.000	Dec, 1995
•LANE Servers Management Spec v1.0	af-lane-0057.000	Mar, 1996
•LANE v2.0 LUNI Interface	af-lane-0084.000	July, 1997
•Multi-Protocol Over ATM Specification v1.0	af-mpoa-0087.000	July, 1997

Network Management

•Customer Network Management (CNM) for ATM Public Network Service	Af-nm-0019.000	Oct, 1994
•M4 Interface Requirements and Logical MIB	Af-nm-0020.000	Oct, 1994
•CMIP Specification for the M4 Interface	Af-nm-0027.000	Sep, 1995
•M4 Public Network view	af-nm-0058.000	Mar, 1996
•M4 "NE View"	af-nm-0071.000	Jan, 1997
•Circuit Emulation Service Interworking Requirements, Logical and CMIP MIB	af-nm-0072.000	Jan, 1997
•M4 Network View CMIP MIB Spec v1.0	af-nm-0073.000	Jan, 1997
•M4 Network View Requirements & Logical MIB Addendum	af-nm-0074.000	
•ATM Remote Monitoring SNMP MIB	af-nm-test-0080.000	July, 1997

Physical Layer

| •Issued as part of UNI 3.1: | af-uni-0010.002 | |

- 44.736 DS3 Mbps Physical Layer
- 100 Mbps Multimode Fiber Interface
  Physical Layer
- 155.52 Mbps SONET STS-3c
  Physical Layer
- 155.52 Mbps Physical Layer

•ATM Physical Medium Dependent Interface Specification for 155 Mb/s over Twisted Pair Cable	af-phy-0015.000	Sep, 1994
•DS1 Physical Layer Specification	af-phy-0016.000	Sep, 1994
•Utopia	af-phy-0017.000	Mar, 1994
•Mid-range Physical Layer Specification for Category 3 UTP	af-phy-0018.000	Sep, 1994
•6,312 Kbps UNI Specification	af-phy-0029.000	June, 1995
•E3 UNI	af-phy-0034.000	Aug, 1995
•Utopia Level 2	af-phy-0039.000	June, 1995
•Physical Interface Specification for 25.6 Mb/s over Twisted Pair	af-phy-0040.000	Nov, 1995
•A Cell-based Transmission Convergence Sublayer for Clear Channel Interfaces	af-phy-0043.000	Jan, 1996
•622.08 Mbps Physical Layer	af-phy-0046.000	Jan, 1996
•155.52 Mbps Physical Layer Specification for Category 3 UTP (See also UNI 3.1, af-uni-0010.002)	af-phy-0047.000	
•120 Ohm Addendum to ATM PMD Interface Spec for 155 Mbps over TP	af-phy-0053.000	Jan, 1996
•DS3 Physical Layer Interface Spec	af-phy-0054.000	Mar, 1996
•155 Mbps over MMF Short Wave Length Lasers, Addendum to UNI 3.1	af-phy-0062.000	July, 1996
•WIRE (PMD to TC layers)	af-phy-0063.000	July, 1996
•E-1 Physical Layer Interface Specification	af-phy-0064.000	Sep, 1996
•155 Mbps over Plastic Optical Fiber (POF)	af-phy-0079.000	May, 1997
•Inverse ATM Mux	af-phy-0086.000	July, 1997

P-NNI

•Interim Inter-Switch Signaling Protocol	af-pnni-0026.000	Dec, 1994
•P-NNI V1.0	af-pnni-0055.000	Mar, 1996
•PNNI 1.0 Addendum (soft PVC MIB)	af-pnni-0066.000	Sep, 1996
•PNNI ABR Addendum	af-pnni-0075.000	Jan, 1997
•PNNI v1.0 Errata and PICs	af-pnni-0081.000	July, 1997

Service Aspects and Applications

•Frame UNI	af-saa-0031.000	Sep, 1995
•Circuit Emulation	af-saa-0032.000	Sep, 1995
•Native ATM Services: Semantic Description	af-saa-0048.000	Feb, 1996
•Audio/Visual Multimedia Services: Video on Demand v1.0	af-saa-0049.000	Jan, 1996
•Audio/Visual Multimedia Services: Video on Demand v1.1	af-saa-0049.001	Mar, 1997
•ATM Names Service	af-saa-0069.000	Nov, 1996
•FUNI 2.0	af-saa-0088.000	July, 1997
•Native ATM Services DLPI Addendum Version 1.0	af-saa-dlpi-0091.000	February, 1998

Testing

•ATM Security Framework Version 1.0	af-sec-0096.000	February, 1998

Signaling

•UNI Signaling 4.0	af-sig-0061.000	July, 1996
•Signaling ABR Addendum	af-sig-0076.000	Jan, 1997

Testing

•Introduction to ATM Forum Test Specifications	af-test-0022.000	Dec, 1994
•PICS Proforma for the DS3 Physical Layer Interface	af-test-0023.000	Sep, 1994
•PICS Proforma for the SONET STS-3c Physical Layer Interface	af-test-0024.000	Sep, 1994
•PICS Proforma for the 100 Mbps Multimode Fibre Physical Layer Interface	af-test-0025.000	Sep, 1994
•PICS Proforma for the ATM Layer (UNI 3.0)	af-test-0028.000	Apr, 1995
•Conformance Abstract Test Suite for the ATM Layer for Intermediate Systems (UNI 3.0)	af-test-0030.000	Sep, 1995
•Interoperability Test Suite for the ATM Layer (UNI 3.0)	af-test-0035.000	Apr, 1995

•Interoperability Test Suites for Physical Layer: DS-3, STS-3c, 100 Mbps MMF (TAXI)	af-test-0036.000	Apr, 1995
•PICS Proforma for the DS1 Physical Layer	af-test-0037.000	Apr, 1995
•Conformance Abstract Test Suite for the ATM Layer (End Systems) UNI 3.0	af-test-0041.000	Jan, 1996
•PICS for AAL5 (ITU spec)	af-test-0042.000	Jan, 1996
•PICS Proforma for the 51.84 Mbps Mid-Range PHY Layer Interface	af-test-0044.000	Jan, 1996
•Conformance Abstract Test Suite for the ATM Layer of Intermediate Systems (UNI 3.1)	af-test-0045.000	Jan, 1996
•PICS for the 25.6 Mbps over Twisted Pair Cable (UTP-3) Physical Layer	af-test-0051.000	Mar, 1996
•Conformance Abstract Test Suite for the ATM Adaptation Layer (AAL) Type 5 Common Part (Part 1)	af-test-0052.000	Mar, 1996
•PICS for ATM Layer (UNI 3.1)	af-test-0059.000	July, 1996
•Conformance Abstract Test Suite for the UNI 3.1 ATM Layer of End Systems	af-test-0060.000	June, 1996
•Conformance Abstract Test Suite for the SSCOP Sub-layer (UNI 3.1)	af-test-0067.000	Sep, 1996
•PICS for the 155 Mbps over Twisted Pair Cable (UTP-5/STP-5) Physical Layer	af-test-0070.000	Nov, 1996
•PNNI v1.0 Errata and PICs	af-pnni-0081.000	July, 1997
•PICS for Direct Mapped DS3	af-test-0082.000	July, 1997
•Conformance Abstract Test Suite for Signaling (UNI 3.1) for the Network Side	af-test-0090.000	September, 1997
•ATM Test Access Function (ATAF) Specification Version 1.0	af-test-nm-0094.000	February, 1998
•PICS for Signaling (UNI v3.1) – User Side	af-test-0097.000	April, 1998

Traffic Management

•Traffic Management 4.0	af-tm-0056.000	Apr, 1996
•Traffic Management ABR Addendum	af-tm-0077.000	Jan, 1997

Voice & Telephony over ATM

•Circuit Emulation Service 2.0	af-vtoa-0078.000	Jan, 1997

•Voice and Telephony Over ATM to the Desktop	af-vtoa-0083.000	May, 1997
•(DBCES) Dynamic Bandwidth Utilization in 64 Kbps Time Slot Trunking Over ATM - Using CES	af-vtoa-0085.000	July, 1997
•ATM Trunking Using AAL1 for Narrow Band Services v1.0	af-vtoa-0089.000	July, 1997

User-Network Interface (UNI)

•ATM User-Network Interface Specification V2.0	af-uni-0010.000	June, 1992
•ATM User-Network Interface Specification V3.0	af-uni-0010.001	Sep, 1993
•ATM User-Network Interface Specification V3.1	af-uni-0010.002	1994

C. IETF's ATM-Related RFC Standards

•RFC1483 Multiprotocol Encapsulation over ATM Adaptation Layer
•RFC1577 Classical IP and ARP over ATM
•RFC1755 ATM Signaling Support for IP over ATM
•RFC1626 Default IP MTU for use over ATM AAL5
•RFC1695 Definitions of Managed Objects for ATM Management Version 8.0 using SMIv2

D. ANSI T1 Committee Specifications for ATM

ANSI T1 Standards Related to ATM:

• ANSI T1.105-1995, Synchronous Optical Network (SONET) – Basic Description including Multiplex Structure, Rates and Formats
• ANSI T1.408-1990, ISDN Primary Rate - Customer Installation Metallic Interfaces, Layer 1 Specification
• ANSI T1.511-1994, B-ISDN ATM Layer Cell Transfer-Performance Parameter
• ANSI T1.624-1993, (Superceded by T1.646-1995)
• ANSI T1.627-1993, B-ISDN ATM Layer Functionality and Specification
• ANSI T1.629-1993, B-ISDN ATM Adaptation Layer 3/4 Common Part Functions and Specification
• ANSI T1.630-1993, B-ISDN ATM Adaptation Layer Constant Bit Rate Service Functionality and Specification
• ANSI T1.634-1993, Frame Relay Service Specific Convergence Sublayer

- ANSI T1.635-1994, B-ISDN ATM Adaptation Layer Type 5 Common Part Functions and Specification
- ANSI T1.636-1994, B-ISDN Signaling ATM Adaptation Layer Overview Description
- ANSI T1.637-1994, B-ISDN ATM Adaptation Layer - Service Specific Connection Oriented Protocol (SSCOP)
- ANSI T1.638-1994, B-ISDN Signaling ATM Adaptation Layer - Service Specific Coordination Function for Support of Signaling at the User-To-Network Interface (SSCF at the UNI)
- ANSI T1.640-1996, B-ISDN Network Node Interfaces and Inter Network Interfaces - Rates and Formats Specifications
- ANSI T1.644-1995, B-ISDN - Meta-Signaling Protocol
- ANSI T1.645-1995, B-ISDN Signaling ATM Adaptation Layer - Service Specific Coordination Function for Support of Signaling at the Network Node Interface (SSCF at the NNI)
- ANSI T1.646-1995, B-ISDN Physical Layer Specification for User-Network Interfaces Including DS1/ATM
- ANSI T1.648-1995, B-ISDN User Part (B-ISUP)
- ANSI T1.652-1996, B-ISDN Signaling ATM Adaptation Layer - Layer Management for the SAAL at the NNI
- ANSI T1.654-1996, B-ISDN Operations and Maintenance Principles and Functions
- ANSI T1.656-1996, B-ISDN Interworking between SS7 Broadband ISDN User Part (B-ISUP) and ISDN User Part (ISUP)
- ANSI T1.657-1996, B-ISDN Interworking between SS7 Broadband ISDN User Part (B-ISUP) and Digital Subscriber Signaling System No. 2 (DSS2)
- ANSI T1.658-1996, B-ISDN Extensions to the SS7 B-ISDN User Part, Additional Traffic Parameters for Sustainable Cell Rate (SCR) and Quality of Service (QoS)
- ANSI T1.652-1996, B-ISDN ATM End System Address for Calling and Called Party
- ANSI T1.663-1996, B-ISDN Network Call Correlation Identifier
- ANSI T1.664-1997, B-ISDN Point-to-Multipoint Call/Connection Control
- ANSI T1.665-1997, B-ISDN Overview of B-ISDN NNI Signaling Capability Set 2, Step 1

T1 Technical Reports on ATM

- TR 46 - Transmission Performance Standards Issues on ATM Voice-Based Telecommunications Network (April 1996)

• TR 53 - Transmission Performance Guidelines for ATM Technology
Intended for Integration into Networks Supporting Voiceband
Services

E. Frame Relay Forum Specifications for ATM

Relevant Frame Relay Forum Specifications:

• FRF.5, Frame Relay/ATM PVC Network Interworking Implementation
Agreement
• FRF.8, Frame Relay/ATM PVC Service Interworking Implementation
Agreement
• FRF.11,Voice over Frame Relay Implementation Agreement

II. ONGOING ATM STANDARDIZATION ACTIVITIES

A. Current Standardization Activities in ITU-T

• E.459 (03/98) Criteria and procedures for the reservation, assignment and
reclamation of E.164 country codes and associated
Identification Codes (ICs)
• E.728 (03/98) Grade of service parameters for B-ISDN signaling

B. Current Standardization Activities in ATM Forum (as of February 1998)

Control Signaling

• ATM Inter-Network Interface (AINI), Security Addendum (12/98)
• UNI 4.1 Signaling (4/99)
• GSS/B-QSIG Interworking PNNI 1.0 Addendum (9/98)

Joint CS and RA

• Interworking Among ATM Networks, PNNI 2.0 (4/99)

Lan Emulation/MPOA

• LANE v2.0 LEC MIB (7/98), LANE v2.0 Server-to-server Interface (2/99)
• Multi-Protocol Over ATM v1.0 MIB (7/98)

Network Management

- Enterprise/Carrier Management Interface (M4) Requirements & Logical MIB SVC Function NE View V2.0
- Enterprise/Carrier Network Management (M4) SNMP MIB (7/98)
- Carrier Interface (M5) Requirements & CMIP MIB
- Management System Network Interface Security Requirements & Logical MIB
- ATM Access Function Specification Requirements & Logical MIB
- M4 Requirements & Logical MIB
- Network View v2.0

Physical Layer

- nxDS0 Interface, 2.4 Gbps Interface (1998)
- 1-2.5 Gbps Interface (1998)
- 10 Gbps Interface
- Addendum to POF af-phy-pof155-0079.000 for Fiber Jack Connector (4/98)
- Utopia Level 3 (1999)
- Inverse Multiplexing for ATM Spec. v1.1

RBB (Residential Broadband)

- RBB Architectural Framework (Straw Ballot 7/98)
- RBB Interfaces Specification (Straw Ballot 7/98)

Routing and Addressing

- PNNI Augmented Routing (PAR) (Work in Progress 9/98)
- ATM Forum Addressing Reference Guide (Work in Progress 9/98)
- ATM Forum Addressing User Guide (Work in Progress 12/98)
- New Capabilities ATM Addressing (Work in Progress TBD)

Security

- ATM Security Specification v1.0 (Straw Ballot 7/98)

Service Aspects & Applications

- API Semantic Doc 2.0 for UNI 4.0 (Work in Progress 11/98)
- H.323 Media Transport over ATM (Work in Progress TBD)
- FUNI Extension for Multi-Media (Work in Progress TBD)
- Native ATM Connectionless Requirements (Work in Progress TBD)
- Java API (Work in Progress TBD)
- Voice to desktop over ATM v2.0 (Work in Progress TBD)

Testing

- Conformance Abstract Test Suite for Signaling (UNI 3.1) for the User Side (Work in Progress 7/98)
- Performance Testing Specification (Work in Progress 7/98)
- PICS for Signaling (UNI v3.1) - User Side (Final Ballot 4/98)
- Conformance Abstract Test Suite for LANE 1.0 Server (Work in Progress TBD)
- Conformance Abstract Test Suite for UNI 3.0/3.1 ILMI Registration (User Side & Network Side) (Work in Progress TBD)
- UNI Signaling Performance Test Suite (Work in Progress TBD)
- Interoperability Test Suite for PNNI v1.0 (Work in Progress 11/98)
- Interoperability Test Suite for LANE v1.0 (Work in Progress TBD)
- PICS for Signaling (UNI v3.1) Network Side (Work in Progress TBD)
- Introduction to ATM Forum Test Specification v2.0 (Work in Progress 9/98)
- Conformance Abstract Test Suite for SSCOP v2.0 (Straw Ballot 7/98)

Traffic Management

- Traffic Management 5.0 (Work in Progress 12/98)

Voice and Telephony over ATM

- ATM Trunking Using AAL2 for Narrowband (Work in Progress TBD)
- Low Speed CES (Work in Progress TBD)

Wireless ATM

WATM Spec 1.0 (Work in Progress April / May 1999)

C. IETF's Current Standardization Activities

Draft Standard

• RFC2226 IP Broadcast over ATM Networks

REFERENCES

[Aras 1994] C. M. Aras, J. F. Kurose, D. S. Revees, and H. Schulzrinne, Proceedings of the IEEE, "Real-Time Communication in Packet Switch Networks", vol. 82, no. 1, pp. 122-138, January 1994

[Black 1997] U.D. Black, Signaling in Broadband Networks, Prentice Hall, Englewood Cliffs, NJ, vol. 2, 1997

[Brown 1989] T. X. Brown, IEEE Communications, "Neural Networks for Switching", pp. 72-81, November 1989

[Brown 1990] T. X. Brown and K. Liu, IEEE Journal on Selected Areas in Communications, "Neural Network Design of a Banyan Network Controller", vol. 8, no. 3, pp. 1428-1438, October 1990

[Burg 1991] J. Burg and D. Dorman, IEEE Communications Magazine, "Broadband ISDN Resource Management: The Role of Virtual Paths", September 1991

[Chang 1986] F. Chang and L. Wu, IEEE Transactions on Automata Control, "An Optimal Adaptive Routing Algorithm", vol. ac-31, no. 8, pp. 690-700, August 1986

[Chao 1992] H. Jonathan Chao, in Proc. ACM SIGCOMM'92, "Architecture Design for Regulating and Scheduling User's Traffic in ATM Networks", Baltimore, MD, pp. 77-87, August 1992

[Cidon 1988] I. Cidon, I. Gopal, G. Grover, and M. Sidi, IEEE Journal on Selected Areas in Communications, "Real-Time Packet Switching: A Performance Analysis", vol. 6, no. 9, pp. 1576-1586, December 1988

[Cooper 1981] R. B. Cooper, Introduction to Queuing Theory, 2nd Edition, Elsevier Science Publishing Co., New York, New York, 1981

[Cooper 1990] C. A. Cooper and K. Park, IEEE Network Magazine, "Towards a Broadband Congestion Control Strategy", pp. 18-23, May 1990

[Gerla 1990] M. Gerla, L. Fratta, and J. Monterio, in Proc. SBT/IEEE International Telecommunications Symposium, "Leaky Bucket Analysis for ATM Networks", Rio de Janerio, Brazil, pp. 482-502, September 1990

[Ginsberg, 1996] D. Ginsberg, ATM: solutions for enterprise networking, Addison-Wesley, Reading, MA, 1996

[Goralski, 1995] W.J.Goralski, Introduction to ATM Networking, McGraw-Hill, New York, 1995

[Hiramatsu 1990] A. Hiramatsu, IEEE Transactions on Neural Networks, "ATM Communications Network Control by Neural Networks", vol. 1, no. 1, pp. 121-130, March 1990

[Hiramatsu 1991] A. Hiramatsu, IEEE Journal on Selected Areas in Communications, "Integration of ATM Call Admission Control and Link Capacity Control by Distributed Neural Networks", vol. 9, no. 7, pp. 1131-1138, September 1991

[Hopfield 1985] J. J. Hopfield and D. W. Tank, Biological Cybernetics, " Neural Computation of Decisions in Optimization Problems", vol. 52, pp. 141-152, 1985

[Hui 1987] J. Hui and E. Arthurs, IEEE Journal on Selected Areas in Communications, "A Broadband Packet Switch for Integrated Transport", vol. SAC-5, no. 8, pp. 1264-1273, October 1987

[Hui 1988] J. Y. Hui, IEEE Journal on Selected Areas in Communications, "Resource Allocation for Broadband Networks", vol. 6, no. 9, pp. 1598-1608, December 1988

[Jain 1990] R. Jain, IEEE Network Magazine, "Congestion Control in Computer Networks: Issues and Trends", pp. 24-30, May 1990

[Jain 1991] R. Jain, The Art of Computer Systems Performance Analysis, John Wiley & Sons, 1991

[Jenq 1983] Y. C. Jenq, IEEE Journal on Selected Areas in Communications, "Performance Analysis of a Packet Switch Based on Single-Buffered Banyan Network", vol. sac-1, no. 6, pp. 1014-1021, December 1983

[Kamoun 1980] F. Kamoun and L. Kleinrock, IEEE Transactions on Communications, "Analysis of Shared Finite Storage in a Computer Node Environment Under General Traffic Conditions", vol. com-28, no. 7, pp. 992-1003, July 1980

[Karol 1987] M. J. Karol, M. G. Hluchyj, and S. P. Morgan, IEEE Transactions on Communications, "Input Versus Output Queueing on a Space-Division Packet Switch", vol. com-35, no. 12, pp. 1347-1356, December 1987

[Karol 1988] M. J. Karol and M. G. Hluchyj, IEEE Journal on Selected Areas in Communications, "Queueing in High-performance Packet Switching", vol. 6, no. 9, pp. 1587-1597, December 1988

[Katevenis 1987] M. G. H. Katevenis, IEEE Journal on Selected Areas in Communications, "Fast Switching and Fair Control of Congested Flow in Broadband Networks", vol. sac-5, no. 8, pp. 1315-1326, October 1987

[Khosrow 1991] K. Khosrow and M. Sidi, in Proc. IEEE INFOCOM'91, "On the Performance of Bursty and Correlated Sources to Leaky Bucket Rate-Based Access Control Schemes", pp. 426-434, April, 1991

[Kim 1990] H. S. Kim and A. Leon-Garcia, IEEE Transactions on Communications, "Performance of Buffered Banyan Networks Under Nonuniform Traffic Patterns", vol. 38, no. 5, May 1990

[Lang 1990] T. Lang and L. Kurisaki, Journal of Parallel and Distributed Computing, "Nonuniform Traffic Spots (NUTS) in Multistage Interconnections Networks", no. 10, pp. 55-57, 1990

[Leduc 1994] J. P. Leduc, Digital Moving Pictures - Coding and Transmission on ATM Networks, Advances in Image Communication Series, Elsevier Science B.V., Amsterdam, The Netherlands, 1994

[Lee 1993] S. Lee and S. Chang, IEEE Transactions on Neural Networks, "Neural Networks for Routing of Communication Networks with Unreliable Components", vol. 4, no. 5, September 1993

[McDysan 1995] D. E. McDysan and D. L Spohn, ATM Theory and Application, McGraw-Hill Series on Computer Communications, McGraw-Hill Inc., 1995

[Mun 1994] Y. Mun and H. Y. Yong, IEEE Transactions on Computers, "Performance Analysis of a Finite Buffered Multistage Interconnection Networks", vol. 43, no. 2, pp. 153-161, February 1994

[Muralidar 1987] K. H. Muralidar and M. K. Sundareshan, IEEE Transactions on Automatic Control, "Combined Routing and Flow Control in Computer Communication Networks: A Two-Level Adaptive Scheme", vol. ac-32, no. 1, pp. 15-25, January 1987

[Pandya 1995] A.S. Pandya and R. B. Macy, Pattern Recognition with Neural Networks in C++, CRC Press, Boca Raton and IEEE Press, 1995

[Rauch 1988] H. E. Rauch and T. Winarske, IEEE Control Systems Magazine, "Neural Networks for Routing Communication Traffic", pp. 26-31, April 1988

[Re 1993] E. D. Re and R. Fantacci, IEE Proceedings-I, "Efficient Fast Packet Switch Fabric with Shared Input Buffers", vol. 140, no. 5, pp. 372-380, October 1993

[Rodriguez 1990] M. A. Rodriguez, IEEE Network Magazine, "Evaluating Performance of High-Speed Multiaccess Networks", pp. 36-41, May 1990

[Sen 1995] E. Sen, A.S. Pandya and S. Hsu, in Proc. SPIE Vol. 2492, "Neural Network Based Buffer Allocation Scheme for ATM Networks", pp. 51-57, Applications and Science of Artificial Neural Networks, S.K. Rogers; D.W. Ruck; Eds., April 1995

[Stallings 1995] W. Stallings, ISDN and Broadband ISDN with Frame Relay and ATM, Prentice Hall, Englewood Cliffs, NJ, 1995

[Tanner 1995] M. Tanner, Practical Queuing Analysis, The IBM McGraw-Hill Series, McGraw-Hill Inc., 1995

[Troudet 1991] T. P. Troudet and S. M. Walters, IEEE Transactions on Circuits and Systems, "Neural network architecture for crossbar switch control", vol. 38, no. 1, pp. 42-56, January 1991

ATM Acronyms

AAL - ATM Adaptation Layer
ABR - Available Bit Rate
ACR- Adaptive Clock Recovery
ADSL - Asymmetrical Digital Subscriber Line
ANSI - American National Standards Institute
API - Application Programming Interface
ARP - Address Resolution Protocol
ASIC - Application Specific Integrated Circuit
ATM - Asynchronous Transfer Mode
ATMARP - ATM Address Resolution Protocol
ASM - ATM Switching Matrix
ASP-down - ATM Switching Preprocessor Downstream
ASP-up - ATM Switching Preprocessor Upstream
B-ICI - Broadband Inter Carrier Interface
B-ISDN - Broadband Integrated Services Digital Network
B-ISUP - Broadband ISDN User's Part
B-LLI - Broadband Low Layer Information
B-NT - Broadband Network Termination
B-TE - Broadband Terminal Equipment
BECN - Backward Explicit Congestion Notification
BER - Bit Error Rate
BISDN - Broadband - Integrated Services Digital Network
BPS - Bits per second
BT - Burst Tolerance
BUS - Broadcast and Unknown Server
BW - Bandwidth
CAC - Connection Admission Control
CAS - Channel Associated Signaling
CBR - Constant Bit Rate
CCITT - Consultative Committee on International Telephone & Telegraph
CCR - Current Cell Rate
CCS - Common Channel Signaling
CCSS7 - Common Channel Signaling System 7

CDT - Cell Delay Tolerance
CDV - Cell Delay Variation
CDVT - Cell Delay Variation Tolerance
CEI - Connection Endpoint Identifier
CER - Cell Error Ratio
CES - Circuit Emulation Service
CES-IS - CES Interworking Service
CES-IWF - CES Interworking Function
CI - Congestion Indicator
CIP - Carrier Identification Parameter
CIR - Committed Information Rate
CLEC - Competitive Local Exchange Carrier
CLP - Cell Loss Priority
CLIP - Classical IP over ATM
CLR - Cell Loss Ratio
CMIP - Common Management Interface Protocol
CNM - Customer Network Management
CO - Central Office
COS - Class of Service
CP - Complete Partitioning
CPCS - Common Part Convergence Sublayer
CPE - Customer Premises Equipment
CPI - Common Part Indicator
CPN - Customer Premises Network
CPN - Calling Party Number
CRC - Cyclic Redundancy Check
CS - Complete Sharing
CS - Convergence Sublayer
CSI - Convergence Sublayer Indication
CSR - Cell Missequenced Ratio
CSU - Channel Service Unit
CTC - Cell Transfer Capacity
CTD - Cell Transfer Delay
CTV - Cell Tolerance Variation
DA - Destination MAC Address
DBS - Direct Satellite Broadcast
DCC - Data Country Code
DCE - Data Communication Equipment
DES - Destination End System
DLC - Data Link Control
DLCI - Data Link Connection Identifier
DLL - Data Link Layer
DLPI - Data Link Provider Interface
DQDB - Distributed Queue Dual Bus
DS0 - Digital Signal, Level 0

DS1 - Digital Signal, Level 1
DS3/DS-3 - Digital Signal, Level 3
DS3 PLCP - Physical Layer Convergence Protocol
DSID - Destination Signaling Identifier
DSLAM - Digital Subscriber Line Access Multiplexer
DSS1 - Digital Subscriber Signaling #1
DSS2 - Digital Subscriber Signaling #2
DSU - Data Service Unit
DTE - Data Terminal Equipment
DWDM - Dense Wavelength Division Multiplexing
DXI - Data Exchange Interface
EFCI - Explicit Forward Congestion Indication
EKTS - Electronic Key Telephone System
ELAN - Emulated Local Area Network
ESF - Extended Super Frame
ESI - End-System Identifier
FCS - Frame Check Sequence
FDDI - Fiber Distributed Data Interface
FDL - Facility Data Link
FEBE - Far End Block Error
FEC - Forward Error Correction
FERF - Far End Receive Failure
FIFO - First-In First-Out
FR - Frame Relay
FRS - Frame Relay Service
FUNI - Frame User Network Interface
GCRA - Generic Cell Rate Algorithm
GFC - Generic Flow Control
HDB3 - High Density Bipolar 3
HDLC - High Level Data Link Control
HDSL - High-speed Digital Subscriber Line (symmetrical)
HDTV - High Definition Television
HEC - Header Error Check
HIPPI - High Performance Parallel Interface
HOL - Head of Line
IEC - Inter-exchange Carrier
IEEE - Institute of Electrical and Electronics Engineers
IETF - Internet Engineering Task Force
ILEC - Incumbent Local Exchange Carrier
ILMI - Interim Local Management Interface
IMA - Inverse Multiplexing over ATM
IP - Internet Protocol
IPNG - Internet Protocol Next Generation
IPX - Novell Internetwork Packet Exchange
ISO - International Organization for Standardization

ISP - Internet Service Provider
ITU-T - International Telecommunications Union - Telecommunication
IWF - Interworking Function
IWU - Interworking Unit
JPEG - Joint Photographic Experts Group
LAN - Local Area Network
LANE - Local Area Network Emulation
LAPD - Link Access Procedure D
LCT - Last Conformance Time
LD - LAN Destination
LE - LAN Emulation
LE_ARP - LAN Emulation Address Resolution Protocol
LEC - Local Exchange Carrier
LEC - LAN Emulation Client
LECID - LAN Emulation Client Identifier
LECS - LAN Emulation Configuration Server
LES - LAN Emulation Server
LIM - Line Interface Module
LIS - Logical IP Subnetwork
LLC - Logical Link Control
LLC/SNAP - Logical Link Control/Subnetwork Access Protocol
LMI - Layer Management Interface
LOC - Loss of Cell delineation
LOF - Loss of Frame
LOS - Loss of Signal
MAC - Medium Access Control
MAN - Metropolitan Area Network
MARS - Multicast Address Resolution Server
MBS - Maximum Burst Size
MCR - Minimum Cell Rate
MCS - Multicast Server
MCTD - Mean Cell Transfer Delay
MIB - Management Information Base
MIN - Multistage Interconnection Networks
MIR - Maximum Information Rate
MMBP(2) - Markov Modulated Bernoulli Process with two states
MMF - Multimode Fiberoptic cable
MPC - MPOA Client
MPEG - Motion Picture Experts Group
MPOA - Multiple Protocol over ATM
MPS - MPOA Server
MSAP - Management Service Access Point
N-ISDN - Narrowband Integrated Services Digital Network
NC - Network Centric
NDIS - Network Driver Interface Specification

NE - Network Element
NEBIOS - Network Basic Input/Output System
NHC - NHRP Client
NHRP - Next Hop Resolution Protocol
NHS - NHRP Server
NMS - Network Management System
NNI - Network to Network Interface
NP - Non Polynomial
NPC - Network Parameter Control
NRM - Network Resource Management
NSAP - Network Service Access Point
NSP - Network Service Provider
NSR - Non-Source Routed
NT - Network Termination
OAM - Operations, Administration and Maintenance
OC-3c - Optical Carrier, Level 3 Concatenated (155 Mbit/s)
OC-12c - Optical Carrier, Level 12 Concatenated (622 Mbit/s)
OC-48c - Optical Carrier, Level 48 Concatenated (2.4 Gbit/s)
OC-192 - Optical Carrier, Level 192 (9.6 Gbit/s)
ODI - Open Data-Link Interface
OLI - Originating Line Information
OOF - Out of Frame
OSI - Open Systems Interconnection
OSID - Origination Signaling Identifier
OSPF - Open Shortest Path First
P-NNI - Private Network to Network Interface
PAD - Packet Assembler and Disassembler
PBX - Private Branch eXchange
PC - Priority Control
PCM - Pulse Code Modulation
PCR - Peak Cell Rate
PCR - Program Clock Reference
PCVS - Point to Point Switched Virtual Connections
PD - Packetization Delay
PDH - Plesiochronous Digital Hierarchy
PDN - Public Data Networks
PDU - Protocol Data Unit
PHY - Physical Layer
PLL - Phase Locked Loop
PLPC - Physical Layer Convergence Protocol
PM - Physical Medium
PMD - Physical Layer Dependent sub-layer
PMD - Polarization Mode Dispersion
PN - Private Network
POH - Path Overhead

POI - Path Overhead Indicator
POT - Plain Old Telephony
PRS - Primary Reference Source
PS - Partial Sharing
PSS1 - Private Subscriber Signaling System #1
PSTN - Public Switched Telephone Network
PT - Payload Type
PTI - Payload Type Identifier
PVC - Permanent Virtual Circuit
PVCC - Permanent Virtual Channel Connection
PVPC - Permanent Virtual Path Connection
QD - Queuing Delay
QoS - Quality of Service
QPSX - Queue Packet and Synchronous Circuit Exchange
RAI - Remote Alarm Indication
RBOC - Regional Bell Operating Company
RC - Routing Control
RD - Route Descriptor
RDF - Rate Decrease Factor
RDI - Remote Defect Identification
RDI - Remote Defect Indication
RFC - Request For Comment (Document Series)
RI - Routing Information
RII - Routing Information Indicator
RIP - Routing Information Protocol
RM - Resource Management
RSVP (protocol) - Resource Reservation Protocol
RTS - Residual Time Stamp
SAAL - Signaling ATM Adaptation Layer
SAP - Service Access Point
SAR - Segmentation and Reassembly
SC - Sequence Counter
SCCP - Signaling Connection and Control Part
SCP - Service Control Point
SCR - Sustainable Cell Rate
SDH - Synchronous Digital Hierarchy
SDT - Structured Data Transfer
SDU - Service Data Unit
SE - Switching Element
SF - Super Frame
SID - Signaling Identifier
SIPP - SMDS Interface Protocol
SIR - Sustained Information Rate
SLIF - Switch Link Interface
SMA - Sharing with a Minimum Allocation

SMDS - Switched Multi-megabit Data Services
SMF - Single Mode Fiber
SMQMA - Sharing with a Maximum Queue and Minimum Allocation
SMXQ - Sharing with Maximum Queue Lengths
SN - Sequence Number
SNA - Systems Network Architecture
SNAP - Sub Network Access Protocol
SNMP - Simple Network Management Protocol
SNP - Sequence Number Protection
SOH - Section Overhead
SONET - Synchronous Optical Network
SPID - Service Protocol Identifier
SR - Source Routing (Bridging)
SRF - Specifically Routed Frame
SRT - Source Routing Transparent
SRTS - Synchronous Residual Time Stamp
SSCF - Service Specific Coordination Function
SSCOP - Service Specific Connection Oriented Protocol
SSCS - Service Specific Convergence Sublayer
STM - Synchronous Transfer Mode
STM1 - Synchronous Transport Mode 1
STS-3c - Synchronous Transport System-Level 3 concatenated
SVC - Switched Virtual Circuit
SVCI - Switched Virtual Circuit Identifier
SVP - Switched Virtual Path
T1S1 - ANSI T1 Subcommittee
TAT - Theoretical Arrival Time
TC - Transmission Convergence
TCP/IP - Transmission Control Program/Internet Protocol
TCS - Transmission Convergence Sublayer
TDM - Time Division Multiplexing
TE - Terminal Equipment
TM - Traffic Management
TS - Traffic Shaping
TV - Televison
UBR - Unspecified Bit Rate
UDP - User Datagram Protocol
UNI - User Network Interface
UPC - Usage Parameter Control
UTOPIA - Universal Test & Operations PHY Interface for ATM
UTP - Unshielded Twisted Pair cable
VBR - Variable Bit Rate
VC - Virtual Channel
VC-SW - Virtual Channel Switching
VCC - Virtual Channel Connection

VCI - Virtual Circuit Identifier
VCI - Virtual Connection Identifier
VCI - Virtual Channel Identifier
VDSL - Very high-speed Digital Subscriber Line
VLAN - Virtual Local Area Network
VOD - Video on Demand
VOI - Voice over IP
VP - Virtual Path
VP-Sw - Virtual Path Switching
VP/VC - Virtual Path, Virtual Circuit
VPC - Virtual Path Connection
VPI - Virtual Path Identifier
VPI/VCI - Virtual Path Identifier/Virtual Channel Identifier
VPN- Virtual Private Network
VTOA - Voice Telephony over ATM
WAN - Wide Area Network
xDSL- various Digital Subscriber Line

Glossary

A

AAL

ATM Adaptation Layer: The ATM standards layer that allows multiple applications to have data converted to and from the ATM cell. A protocol is used to translate higher layer services into the size and format of an ATM cell.

AAL-1

ATM Adaptation Layer Type 1: AAL functions that are needed for supporting constant bit rate, time-dependent traffic such as voice and video.

AAL-2

ATM Adaptation Layer Type 2: This AAL is yet to be defined by the International Standards bodies. It can be viewed as a placeholder for variable bit rate video transmission.

AAL-3/4

ATM Adaptation Layer Type 3/4: AAL functions for supporting variable bit rate, delay-tolerant data traffic requiring some sequencing and/or error detection support. Originally there were two AAL types, i.e., connection-oriented and connectionless, that were combined to form this layer.

AAL-5

ATM Adaptation Layer Type 5: AAL functions for supporting variable bit rate, delay-tolerant connection-oriented data traffic requiring minimal sequencing or error detection support.

ABR

Available Bit Rate: A type of traffic for which the ATM network attempts to meet that traffic's bandwidth requirements. It does not guarantee a specific amount of bandwidth and the end station must retransmit any information that did not reach the far end.

Address Resolution

Address Resolution is the procedure by which a client associates a LAN destination with the ATM address of another client or the BUS.

AIS

Alarm Indication Signal: An 'all ones' signal sent down or up stream by a device when it detects an error condition or receives an error condition or receives an error notification from another unit in the transmission path.

AMI

Alternate Mark Inversion: A line coding format used on T1 facilities that transmits ones by alternate positive and negative pulses.

ANSI

American National Standards Institute: A U.S. standards body.

API

Application Program Interface: API is a standard programming interface used for communications between software programs or for interfacing between adjacent protocol layers. API_connection is a relationship between an API_endpoint and other ATM devices.

ARP

Address Resolution Protocol: The procedures and messages in a communications protocol which determines which physical network address (MAC) corresponds to the IP address in the packet.

ATM

Asynchronous Transfer Mode: A transfer mode in which the information is organized into cells. It is asynchronous in the sense that the recurrence of cells containing information from an individual user is not necessarily periodic.

ATM Address

Defined in the UNI Specification as 3 formats, each having 20 bytes in length including country, area and end-system identifiers.

ATM Layer Link

A section of an ATM Layer connection between two adjacent active ATM Layer entities.

ATM Link

A virtual path link (VPL) or a virtual channel link (VCL).

ATM Traffic Descriptor

A generic list of traffic parameters that can be used to capture the intrinsic traffic characteristics of a requested ATM connection.

B

B-ICI

Broadband ISDN Inter-Carrier Interface: An ATM Forum defined specification for the interface between public ATM networks to support user services across multiple public carriers.

B-ISDN

Broadband ISDN: A high-speed network standard (above 1.544 Mbps) that evolved narrowband ISDN with existing and new services with voice, data and video in the same network.

B-TE

Broadband Terminal Equipment: An equipment category for B-ISDN which includes terminal adapters and terminals.

BER

Bit Error Rate: A measure of transmission quality.

B-ISUP

Broadband ISDN User's Part: A SS7 protocol which defines the signaling messages to control connections and services.

BT

Burst Tolerance: BT applies to ATM connections supporting VBR services and is the limit parameter of the GCRA.

BUS

Broadcast and Unknown Server: This server handles data sent by an LE Client to the broadcast MAC address ('FFFFFFFFFFFF'), all multicast traffic, and initial unicast frames which are sent by a LAN Emulation Client.

BW

Bandwidth: A numerical measurement of throughput of a system or network.

C

CAC

Connection Admission Control: the set of actions taken by the network during either the call setup phase or the call re-negotiation phase, in order to determine whether a connection request can be accepted/rejected or whether a request for re-allocation can be accommodated.

CAS

Channel Associated Signaling: A form of circuit state signaling in which the circuit state is indicated by one or more bits of signaling status sent repetitively and associated with that specific circuit.

CBR

Constant Bit Rate: An ATM service category which supports a constant or guaranteed rate to transport services. Video or voice as well as circuit emulation which require rigorous timing control and performance parameters are examples of such applications.

CCS

Common Channel Signaling: Voice signaling in which a group of circuits share a signaling channel.

CDV

Cell Delay Variation: CDV is a component of cell transfer delay. It is usually induced by buffering and cell scheduling. Peak-to-peak CDV is a QoS delay parameter associated with CBR and VBR services.

CE

Connection Endpoint: A terminator at one end of a layer connection within a SAP.

CEI

Connection Endpoint Identifier: Identifier of a Connection Endpoint that can be used to identify the connection at a SAP.

Cell

The basic information unit of transmission in ATM. It is a fixed-size frame which consists of a 5-octet header and a 48-octet payload.

Cell Header

Protocol control information for the ATM Layer.

CER

Cell Error Ratio: The ratio of errored cells to the total cells sent in a transmission. The measurement is taken over a certain time interval. It is desirable that it be measured on an in-service circuit.

CES

Circuit Emulation Service: The ATM Forum circuit emulation service interoperability specification for supporting Constant Bit Rate (CBR) traffic over ATM networks.

CI

Congestion Indicator: This is a field in a RM-cell. It is used to cause the source to decrease its Allowed Cell Rate.

CIR

Committed Information Rate: the information transfer rate which a network is offering a user with frame relay service.

CL

Connectionless Service: A service which allows the transfer of information among service subscribers without the need for end-to-end establishment procedures.

CLP

Cell Loss Priority: A one bit field in the ATM cell header that indicates two levels of priority for ATM cells. CLP=0 cells are higher priority than CLP=1 cells, which are more likely to be discarded by an ATM network experiencing congestion.

CLR

Cell Loss Ratio: It is defined for a connection as Lost Cells/Total Transmitted Cells.

CMIP

Common Management Interface Protocol: An ITU-TSS standard for the message formats and procedures used to exchange management information. It is essential in order to operate, administer, maintain and provision a network.

Connection

An ATM connection consists of a concatenation of ATM Layer links in order to provide an end-to-end information transfer capability to access points.

Connectionless

Refers to ability of existing LANs to send data without previously establishing connections.

CPCS

Common Part Convergence Sublayer: A sublayer of an AAL that remains the same regardless of the traffic type.

CPE

Customer Premises Equipment: A term for equipment that resides on the customer's premise. It may not be owned by the local exchange carrier.

CRC

Cyclic Redundancy Check: A checksum that is used to ensure that the upper-layer data is not corrupted. It is a mathematical algorithm that computes a numerical value based on the bits in a block of data. This number is transmitted with the data and the receiver uses this information and the same algorithm to insure the accurate delivery of data by comparing the results of algorithm and the number received. If a mismatch occurs, an error in transmission is presumed.

CRF
Cell Relay Function: the basic function that an ATM network performs in order to provide a cell relay service to ATM end-stations.

CRF
Connection Related Function: Traffic Management uses this term to refer to either a point in a network or a network element where per connection functions are occurring.

CRS
Cell Relay Service: A carrier service which supports the receipt and transmission of ATM cells between end users in compliance with ATM standards and implementation specifications.

CS
Convergence Sublayer: A sublayer of the AAL where traffic is adapted based on its type. It consists of the general procedures and functions that convert between ATM and non-ATM formats.

CTD
Cell Transfer Delay: It is defined as the average transit delay of cells across a connection.

D

Data Connections
Data VCCs connect the LAN Emulation Configuration Servers to each other and to the Broadcast and Unknown Server. These carry flush messages as well as Ethernet/IEEE 802.3 or IEEE 802.5 data frames.

DCE
Data Communication Equipment: A computing equipment at the network end that attaches to a customer via a Data Terminal Equipment.

Demultiplexing
A function performed by a layer entity that identifies and separates SDUs from a single connection to more than one connection.

DES

Destination End Station: It is a termination point which is the destination for ATM messages of a connection. It is typically used as a reference point for ABR services.

DS-0

Digital Signal, Level 0: The 64 kbps rate that is the basic building block for both the North American and European digital hierarchies.

DS-1

Digital Signal, Level 1: The North American Digital Hierarchy signaling standard for transmission at 1.544 Mbps, supporting 24 simultaneous DS-0 signals. The term is often called T1.

DS-2

Digital Signal, Level 2: The North American Digital Hierarchy signaling standard for transmission of 6.312 Mbps that is used by a T2 carrier which supports 96 calls.

DS-3

Digital Signal, Level 3: The North American Digital Hierarchy signaling standard for transmission at 44.736 Mbps that is used by a T3 carrier. DS-3 supports 28 DS-1s plus overhead.   672 calls

DSU

Data Service Unit: It is the equipment which attaches the users' computing equipment to a public network.

DTE

Data Terminal Equipment: An external networking interface equipment such as a modem which is at the customer end of the circuit.

DXI

Data Exchange Interface: A variable length frame-based ATM interface between a DTE and a special ATM CSU/DSU.

E

E.164

A public network addressing standard used by ITU and ATM. It utilizes up to a maximum of 15 digits.

E-1

The 2.048 Mbps transport rate used by the European CEPT carrier to transmit 30 64 kbps digital channels for voice or data calls, plus a 64 kbps signaling channel and a 64 kbps channel for framing and maintenance.

E-3

The 34.368 Mbps transport rate used by the European CEPT carrier to transmit 16 CEPT1s plus overhead.

Edge Device

A physical device which is capable of forwarding packets between legacy interworking interfaces (e.g., Ethernet, Token Ring, etc.) and ATM interfaces. This is performed based on data-link and network layer information but which does not participate in the running of any network layer routing protocol.

EFCI

Explicit Forward Congestion Indication: A bit in the PTI field of the ATM cell header. A network element in a congested state may set EFCI so that this indication may be examined by the destination end-system.

ELAN

Emulated Local Area Network: The ATM segment of a Virtual LAN. It is a logical network initiated by using the mechanisms defined by LAN Emulation, including ATM and attached traditional end stations.

EMI

Electromagnetic Interference: Equipment used in high speed data systems, including ATM, that generate and transmit signals in the radio frequency range of the electromagnetic spectrum. If sufficient power from these signals escapes the equipment enclosures or transmission media, interference to other equipment or radio services may occur.

EML

Element Management Layer: An abstraction of the functions provided by systems that manage each network element on an individual basis.

ER

Explicit Rate Mode: an RM-cell field used to limit the source Allowed Cell Rate to a specific value. It is a form of feedback used under ABR service.

ESI

End System Identifier: End System is where an ATM connection is terminated or initiated. The ESI is a 6 octet field that distinguishes multiple nodes at the same level in case the lower level peer group is partitioned.

ETSI

European Telecommunications Standards Institute: The primary telecommunications standards organization within Europe.

F

Fairness

Fairness in relation to traffic control is defined as meeting all the agreed quality of service (QoS) requirements. This is achieved by controlling the order of service for all active connections.

FC

Feedback Control: a congestion control mechanism performed by the network and by the end-systems to regulate the traffic submitted on ATM connections according to the state of network elements.

FDDI

Fiber Distributed Data Interface: A 100 Mbps Local Area Network standard that was developed by ANSI that is designed to work on fiber-optic cables, using techniques similar to Token Ring.

FEC

Forward Error Correction: A technique for detection and correction of errors in a digital data stream.

Firewall

A mechanism for providing security by blocking certain traffic.

FRS

Frame-Relay Service: A connection oriented service that is capable of carrying up to 4096 bytes per frame.

G

G.703

ITU-T Recommendation G.703, "Physical/Electrical Characteristics of Hierarchical Digital Interfaces."

G.704

ITU-T Recommendation G.704, "Synchronous Frame Structures Used at Primary and Secondary Hierarchy Levels."

G.804

ITU-T Recommendation G.804, "ATM Cell Mapping into Plesiochronous Digital Hierarchy (PDH)."

GCAC

Generic Connection Admission Control: This is a form of CAC process used by PNNI to determine if a link has potentially enough resources to support a connection.

GCRA

Generic Cell Rate Algorithm: It is a traffic shaping algorithm. The GCRA is used to define conformance with respect to the traffic contract of the connection. For each cell arrival the GCRA determines whether the cell conforms to the traffic contract.

GFC

Generic Flow Control: GFC is a four-bit field in the ATM header which can be used to provide local functions (e.g., flow control).

H

HDLC

High Level Data Link Control: An ITU-TSS link layer protocol standard for point-to-point and multi-point communications.

Header

The first five bytes of an ATM cell containing Protocol control information.

HEC

Header Error Control: Using the one-byte field in the ATM cell header, ATM equipment may check for an error and correct the contents of the header.

I

ICR

Initial Cell Rate: An ABR service parameter, in cells/sec. It is the rate at which a source should send initially and after an idle period.

IEEE

Institute of Electrical and Electronics Engineers: A worldwide standards-making body for the electronics industry.

IEEE 802.3

A Local Area Network protocol suite commonly known as Ethernet.

IEEE 802.5

A Local Area Network protocol suite commonly known as Token Ring. A standard originated by IBM for a token passing ring network that can be configured in a star topology. Versions supported are 4 Mbps and 16 Mbps.

IETF

Internet Engineering Task Force: The organization that provides the coordination of standards and specification development for TCP/IP networking.

ILMI

Interim Local Management Interface: An ATM Forum defined interim specification for network management functions between an end user and a public or private network and between a public network and a private network.

IMA

Inverse Multiplexing over ATM: A device that allows multiple T1 or E1 communications facilities to be combined into a single broadband facility for the transmission of ATM cells.

IP

Internet Protocol: Originally developed by the Department of Defense to support interworking of dissimilar computers across a network.

IPX

Novell Internetwork Packet Exchange: A built-in networking protocol for Novell Netware.

ISO

International Organization for Standardization: An international organization for standardization, based in Geneva, Switzerland, that establishes voluntary standards and promotes the global trade of 90 member countries.

ITU

International Telecommunications Union: (Previously CCITT). ITU-T is an international body of member countries whose task is to define recommendations and standards relating to the international telecommunications industry. The fundamental standards for ATM have been defined and published by the ITU-T.

L

LAN

Local Area Network: A network designed to move data between stations within a campus.

LANE

LAN Emulation: The set of services, functional groups and protocols which provide for the emulation of LANS utilizing ATM as a backbone to allow connectivity among LAN and ATM attached end stations.

Leaky Bucket Algorithm

An informal term for the Generic Cell Rate Algorithm. Leaky Bucket is the term used as an analogous description of the algorithm used for conformance checking of cell flows from a user or network.

LEC

LAN Emulation Client: The entity in end systems which performs data forwarding, address resolution, and other control functions.

LECID

LAN Emulation Client Identifier: This identifier is contained in the LAN Emulation header. It indicates the ID of the ATM host or ATM-LAN bridge.

LES

LAN Emulation Server: A server function within LANE which implements the control coordination function for the Emulated LAN.

Link

An entity that defines a topological relationship (including available transport capacity) between two nodes in different subnetworks.

LNNI

LANE Network to Network Interface: The standardized interface between two LANE domains.

LUNI

LANE User to Network Interface: The standardized interface between a LE client and a LE Server within the LAN emulation service.

M

MAC

Media Access Control: IEEE specifications for the lower half of the data link layer (layer 2) that defines topology dependent access control protocols for IEEE LAN specifications.

MAN

Metropolitan Area Network: A network designed to carry data over an area larger than a campus such as an entire city and its outlying area.

MBS

Maximum Burst Size: In the signaling message, the Burst Tolerance (BT) is conveyed through the MBS which is coded as a number of cells. The BT together with the SCR and the GCRA determine the MBS that may be transmitted at the peak rate and still be in conformance with the GCRA.

MCDV

Maximum Cell Delay Variance: This is the maximum two-point CDV objective across a link or node for the specified service category.

MCLR

Maximum Cell Loss Ratio: This is the maximum ratio of the number of cells that do not make it across the link or node to the total number of cells arriving at the link or node.

MCR

Minimum Cell Rate: An ABR service traffic descriptor, in cells/sec, that is the rate at which the source is always allowed to send.

MCTD

Maximum Cell Transfer Delay: This is the sum of the fixed delay component across the link or node and MCDV.

Metasignaling

ATM Layer Management (LM) process that manages different types of signaling and possibly semipermanent virtual channels (VCs), including the assignment, removal and checking of VCs.

MIB

Management Information Base: A definition of management items for some network component that can be accessed by a network manager. A MIB includes the names of objects it contains and the type of information retained.

MPEG

Motion Picture Experts Group: An ISO Standards group dealing with video and audio compression techniques and mechanisms for multiplexing and synchronizing various media streams.

MPOA

Multiprotocol over ATM: A new ATM standard issued by the ATM Forum to standardize running multiple network layer protocols over ATM.

Multicasting

The transmit operation of a single PDU by a source interface where the PDU reaches a group of one or more destinations.

Multiplexing

A function within a layer that interleaves the information from multiple connections into one connection.

Multipoint Access

User access in which more than one terminal equipment (TE) is supported by a single network termination.

N

NML

Network Management Layer: An abstraction of the functions provided by systems which manage network elements on a collective basis, so as to monitor and control the network end-to-end.

NNI

Network Node Interface: An interface between ATM switches defined as the interface between two network nodes.

NPC

Network Parameter Control: Network Parameter Control is defined as the set of actions taken by the network to monitor and control traffic from the NNI.

NT

Network Termination: Network Termination represents the termination point of a Virtual Channel, Virtual Path, or Virtual Path/Virtual Channel at the UNI.

O

OAM

Operations Administration and Maintenance: A group of network management functions that provide network fault indication, performance information, and data and diagnosis functions.

OSI

Open Systems Interconnection: A seven (7) layer architecture model for communications systems developed by the ISO for the interconnection of data communications systems. Each layer uses and builds on the services provided by those below it.

P

PBX

Private Branch eXchange: It is a device which provides private local voice switching and voice-related services within the private network.

PCM

Pulse Code Modulation: An audio encoding algorithm which encodes the amplitude of a repetitive series of audio samples. This encoding algorithm converts analog voice samples into a digital bit stream.

PCR

Peak Cell Rate: The Peak Cell Rate, in cells/sec, is the cell rate which the source may never exceed.

PDU

Protocol Data Unit: A PDU is a message of a given protocol comprising payload and protocol-specific control information, typically contained in a header.

PHY

Physical Layer: It provides for transmission of cells over a physical medium connecting two ATM devices.

Physical Link

A real link which attaches two switching systems.

PMD

Physical Media Dependent: This sublayer defines the parameters at the lowest level, such as speed of the bits on the media.

PNNI

Private Network-Network Interface: A routing information protocol that enables extremely scalable, full function, dynamic multi-vendor ATM switches to be integrated in the same network.

PVC

Permanent Virtual Circuit: This is a link with a static route defined in advance, usually by manual setup.

PVCC

Permanent Virtual Channel Connection: It is a Virtual Channel Connection (VCC) provisioned through some network management function and left up indefinitely.

PVPC

Permanent Virtual Path Connection: It is a Virtual Path Connection (VPC) provisioned through some network management function and left up indefinitely.

Q

QoS

Quality of Service: It is defined on an end-to-end basis in terms of several attributes of the end-to-end ATM connection. These attributes are Cell Loss Ratio, Cell Transfer Delay and Cell Delay Variation.

S

SA

Source Address: The address from which the message or data originated.

SAR

Segmentation and Reassembly: Method of breaking up arbitrarily sized packets.

SVC

Switched Virtual Circuit: A connection established via signaling, where the user defines the endpoints when the call is initiated.

T

TC

Transmission Convergence: The TC sublayer transforms the flow of cells into a steady flow of bits and bytes for transmission over the physical medium.

TCP

Transmission Control Protocol: Originally developed by the Department of Defense to support interworking of dissimilar computers across a network.

TDM

Time Division Multiplexing: A method in which a transmission facility is multiplexed among a number of channels by allocating the facility to the channels on the basis of time slots.

TE

Terminal Equipment: Terminal equipment represents the endpoint of ATM connection(s) and termination of the various protocols within the connection(s).

U

UBR

Unspecified Bit Rate: UBR is an ATM service category which does not specify traffic related service guarantees.

UNI

User-Network Interface: An interface point between ATM end users and a private ATM switch, or between a private ATM switch and the public carrier ATM network.

V

VBR

Variable Bit Rate: An ATM Forum defined service category which supports variable bit rate data traffic with average and peak traffic parameters.

VC

Virtual Channel: A communications channel that provides for the sequential unidirectional transport of ATM cells.

VCC

Virtual Channel Connection: A concatenation of VCLs that extends between the points where the ATM service users access the ATM layer.

VP

Virtual Path: A unidirectional logical association or bundle of VCs.

VPC

Virtual Path Connection: A concatenation of VPLs between Virtual Path Terminators (VPTs). VPCs are unidirectional.

VTOA

Voice and Telephony Over ATM: The ATM Forum voice and telephony over ATM service interoperability specifications. They address three applications for carrying voice over ATM networks: desktop (or LAN services), trunking (or WAN services), and mobile services.

W

WAN

Wide Area Network: A network which spans a large geographic area relative to an office and campus environment of a LAN (Local Area Network).

Index